Computer Modelling of Concrete Mixtures

Computer Modelling of Concrete Mixtures

J. D. Dewar

London and New York

First published 1999
by E & FN Spon,
11 New Fetter Lane, London EC4P 4EE

Simultaneously published in the USA and Canada
by Routledge
29 West 35th Street, New York, NY 10001

E & FN Spon is an imprint of the Taylor & Francis Group

© 1999 J.D. Dewar

Typeset in 10 on 12 pt Sabon
by Mathematical Composition Setters Ltd, Salisbury
Printed and bound in Great Britain by
TJ International Ltd, Padstow, Cornwall

British Library Cataloguing in Publication Data
A catalogue record for this book is available
from the British Library

Library of Congress Cataloging in Publication Data
Dewar, J.D.
    Computer modelling of concrete mixtures / J.D. Dewar.
    Includes bibliographical references.
    1. Concrete–Computer simulation.   I. Title.
  TA439.D55   1999
  666'.893'0113–dc21   98–55726

ISBN 0–419–23020–3

Contents

Preface

Computer modelling of fresh and hardened concrete is a rapid-growth area in which the author has been privileged to be at the forefront over many years.

Development has not yet peaked and it is the author's hope that this book will stimulate further efforts. To this end, principles, formulae and data are provided freely to stimulate such work.

As is usual with developments, time is needed for their assessment and their adoption. Thus, there is still plenty of scope for users to benefit from the developments in modelling. In particular, there remain opportunities for benefit in the following areas

- Replacing or drastically reducing the number of laboratory trial concretes required to confirm or update relationships, as may be required periodically by Quality Assurance Schemes.
- In conjunction with routine tests of materials, to continuously monitor and update batch data, batch books and relationships.
- To dovetail with the quality control systems for concrete strength to ensure that the need for early action is identified in the time gap before such systems can detect and trigger action.
- As a tool for Quality Assurance Scheme inspectors in the routine checking of the validity of batch books, relationships and designs.
- As a tool for Quality Assurance Inspectors for investigating suspected or alleged infringements of rules.
- Selecting concretes to meet individual client specifications.
- 'What if' exercises for technical and economic comparisons to investigate new materials, comparing different sources of materials and examining effects of modifying proportions.
- Fast changes of materials and batchbooks to enable management decisions to take immediate effect.
- A tool for Consulting Engineers for assessing the potential of concrete to meet particular specifications especially for high performance concretes.
- A tool for Materials Developers in optimising the designs of existing materials and products and developing new applications.
- A tool for Engineers and Technologists investigating concrete problems.

- A tool for Universities, Research Organisations and individual Researchers in the design of concrete research projects.
- A tool for Universities and Training Organisations for post-graduate and under-graduate studies.

It will be observed that this book concentrates on the common properties of concrete, viz. water demand; slump; air content; strength and does not concern itself with the less common properties. The aim has been to provide the basic parameters of modelling allowing for any later customised 'add-ons' for specialised applications. Readers will also note that the methods outlined are based on materials data commonly available to concrete producers and users.

Acknowledgements

The initial impetus for the work came from Dr T.C. Powers' seminal book on fresh concrete, his other published papers and from personal discussion in the US and subsequent correspondence.

Professor P.J.E. Sullivan and Dr C. D'Mello tutored the last three years' research for the author's doctoral thesis and provided valuable encouragement, technical comment and advice on presentation. For their kindness and the ready availability of staff and facilities of the City University the author is much indebted.

Numerous individuals have assisted through discussion, correspondence and the provision of published and unpublished information including Professor C.L. Page and Dr C. Thornton of the University of Aston, Dr G. Lees of Birmingham University, Professor G. Lloedolff of the University of Stellenbosch, Professor G. Wischers of the VDZ in Dusseldorf, Professor E. Sellevold and Dr E. Mortsell of the Technical University of Trondheim, Professor S. Numata of Nishi-Nippon Institute of Technology, Japan, Dr F. de Larrard of LCPC, France, Mr A. Corish of Blue Circle Cement, Mr B.V. Brown and Mr R. Ryle of Readymix UK Ltd.

Mr P. Barnes of Readicrete Ltd and Mr I. Forder of Readymix (Western) provided detailed results of concrete trials and Mr S.J. Martin of Readymix UK Ltd provided valuable help with tests of aggregates and data on mixtures of additions and cement. Mr I. Smith of Fosroc was particularly helpful in answering queries concerning admixtures. Assistance was also provided in this area by the Cement Admixtures Association, in particular by the chairman Mr C. Keeley and the secretary Mr J. Buekett. Mr R. Boulton of Minelco Ltd supplied information on the use of the Theory of Particle Mixtures for heavyweight concrete. Mr L.K.A. Sear of Tarmac Topmix Ltd provided information on strength v w/c at high values of w/c.

Thanks are due to those who over a number of years have assisted the author with encouragement and advice including, Dr T.A. Harrison of the Quarry Products Association, Mr J.M. Uren of North East Slag Cement Ltd, Dr D.D. Higgins of CSMA, Dr L. Cassar of Italcementi, Mr N. Greig of CCS Associates, Mr M. Roberts of C & G Concrete Ltd, Mr D. Wimpenny of Sir William Halcrow and Prs, Mr P.E.D. Howes and Mr J. Platt of ARC Ltd,

Mr J.F. Richardson of Redland Aggregates Ltd, Mr P. Rhodes of Tilcon Ltd, Mr C.M. Reeves of BFS Services and Mr K.W. Day, concrete consultant from Australia.

A number of research projects at Universities, including Dundee, Leeds and Queen Mary and Westfield have utilised earlier or current computerised versions of parts of the Theory of Particle Mixtures.

Several projects for the Advanced Concrete Technology Diploma of the Institute of Concrete Technology have also utilised the Theory of Particle Mixtures, including projects by Mr R. Austin of AWAL Readymix in Bahrain, Mr J. Wood of ARC Ltd, Mr B. Patel of Readymix (Western) Ltd and Mr A.J. Stevenson of Cornwall Highways Laboratory. Mr Austin's project was particularly helpful in extending its use to concretes with plasticising admixtures.

Mr J. Bennett of Questjay Ltd and Mr S. Pavey translated the programming of the author into a user-friendly marketable Mixsim98.

The Centre for Concrete Information of the BCA made light work for the author by tracing and providing copies of recent research papers. In particular the monthly issue of 'Concrete Current Awareness' enabled papers of potential interest to be identified and accessed quickly.

The author acknowledges with special gratitude, the help and forbearance of his wife, Gret, in taking second place to a computer over many years and for the daunting task of collating extracts from the literature.

Notation

Symbol	Units	Definition
$A-F$		Slope change points in voids ratio diagrams
$a-f$		Slope change points in concrete voids ratio diagrams
a	%	Air content
		Suffix $_1$ for entrapped air and $_2$ for entrained air
A, B		Factors for relating strength and water/cement
a, b, c		Empirical factors for water content adjustments due to air entrainment
b		Empirical factor for the z formula
		suffix $_1$, $_2$, $_3$ or $_4$
CJ		Cohesion adjustment factor
d, D	μm or mm	d or D_1 for size of smaller material
		D or D_0 for size of larger material
		mean size calculated on log basis
		suffix $_c$ for cement, $_p$ for powder, $_{fa}$ for fine aggregate and $_{ca}$ for coarse aggregate
δW	l/m^3 or %	Change in water demand
e		Efficiency factor for an addition
E	%	Entrapped air content
f_{cem}	N/mm^2	Cement strength EN 196 mortar prisms
f_{cu}	N/mm^2	Concrete strength BS 1881 standard cubes
$f_{w/c}$	N/mm^2	Concrete strength at a value of w/c
F		Adjustment factor for size ratio r
F		Empirical factors for strength
		suffix $_{age}$ for age and $_{agg}$ for aggregate
F	m^2/kg	Fineness (Blaine)
F	%	Entrained air content associated with plasticiser
F_s		Water adjustment factor for slump
G		Normal cohesion factor
g		Age (days) of strength test
J		Empirical adjustment factor
k		Various empirical constants
LBD	kg/m^3	Loose bulk density
m		Spacing factor for coarse particles e.g. md
n		Proportion of fine material by volume
		suffix $_x$ for point of safe cohesion

continued

Symbol	Units	Definition
$(1-n)$		Proportion of coarse material by volume
n		Slope of Rosin Rammler distribution
P	%	Cumulative per cent passing a sieve size by volume
P, Q		Empirical concrete strength factors
r		Ratio of smaller size/larger size
R		Composite strength factor for air
RD		Relative density
RS	mm	Reference slump
S		Solids content i.e. solids per unit **total** volume
SL	mm	Slump
SC	%	Water content by mass of powder in Vicat test
T	%	Total air content
U		Voids ratio i.e. voids per unit **solid** volume
		suffix $_0$ for larger material and $_1$ for smaller material
		suffix $''$ for effective voids ratio
		suffix $_n$ when fine material has proportion $_n$
		suffix $_x$ for safe cohesion
		suffix $_{jx}$ for adjusted voids ratio for safe cohesion
V		Voids content i.e. voids per unit **total** volume
		suffix $_c$ for cement
		suffix $_w$ for water, $_{wadj}$ for adjustment for slump
		suffix $_{fa}$ for fine aggregate
		suffix $_{ca}$ for coarse aggregate
		suffix $_a$ for air
w/c	-	Free water/cementitious material
WR	l/m^3	Water reduction for air
X	μm or mm	Notional dimension of void
X		Point of safe cohesion in voids ratio diagram for concrete
x		Free water/cement by mass
x_0	μm	Characteristic size of Rosin Rammler distribution
Z or z	μm or mm	Notional width factor for additional voids near coarse particles

1 Introduction

This work originated with an investigation by the author prior to 1983, to develop a Theory of Particle Mixtures, enabling the behaviour of fresh concrete to be predicted and economic designs to be prepared from a knowledge of readily available properties of the component materials. The results were published in Dewar (1983) and further work was published in Dewar (1986–1998).

The original work relied on the investigations of Powers (1966 and earlier) for introducing the benefits of voids ratio diagrams for demonstrating the behaviour of mixtures. By analysis of such diagrams for aggregates and the development of suitable models, the author was able to develop a tentative Theory of Particle Mixtures and to apply it to aggregates, mortars and concretes. Experience since 1983 identified potential areas for modifications and further development. From 1994 to 1997 extensive research was undertaken by the author as part of a PhD thesis at City University, London designed to achieve the following

- Confirm or modify the basic concepts of the theory, and the empirical constants used in the formulae from more detailed experimental work by the author and others
- Develop the theory and formulae to cover more accurately size ratios between 0.15 and 1
- Appraise published and unpublished research and other technical information, in particular that since 1983
- Investigate and where appropriate take account of the influences of
  - different energy levels of compaction
  - cement properties
  - inclusion of plasticising admixtures and air-entraining agents
  - inclusion of additions such as fly-ash, ground granulated slag and silica fume
- Develop empirical formulae for including entrained air content and strength of concrete within the prediction system.
- Compare the results of using the theory with data and experience from practice

- Develop a method for optimising multi-component mixtures for minimum voids ratio

Each of these aims has been achieved to a greater or lesser extent. Where uncertainties remain they have been identified for future workers. A copy of the thesis is lodged with the library of City University.

This present volume summarises the results of this research undertaken by the author together with the following recent additions

- References obtained during 1998
- Slump v water content relationships
- Modelling of strength v age relationships
- Expansion of the concept of design of component materials
- Examples of computer screen displays from Mixsim98

1.1 Modelling principles and techniques

Hansen (1986) warned that a

> model is merely a theoretical description proposed to explain observed experimental facts and to provide additional insight into the behaviour of a material. A model cannot, and should not, be considered to be a correct description in any absolute sense. This is why more than one model may adequately interpret the known facts.

Beeby (1991) pointed out the dangers of modelling based on empirical curve fitting without a sound physical basis, albeit recognising that an engineer must be willing to employ empiricism and theory while understanding the limitations of both. Beeby considers that some engineers seem to over-value experience, empirical knowledge and rules of thumb and to underrate theory, while others tend to the opposite fault.

Beeby also stresses the relative reliability of empirical methods for interpolation but unreliability for extrapolation and that only a fundamental understanding on the basis of sound theory provides confidence for moving forward into new areas.

The author subscribes to the views of Hansen and Beeby and has attempted, wherever possible, to develop a coherent theory for aggregates, mortars and for fresh concrete, and only to use empirical factors as a last resort. Thus, for example, while the interactions between size and voids ratio have been considered from a theoretical view point, the numerical characterisation of change points in voids ratio diagrams has been dealt with empirically, but with an attempt at a theoretical explanation for some of the empirical values concerning spacing of particles.

A deliberate self-imposed constraint was adopted of using test methods and data commonly available to the concrete industry. This has meant that for modelling of strength the author was reliant upon well-established empirical models, but again with useful aids from theoretical physical concepts, as will be apparent from the discussion in Section 5.4.

Hansen (1986) identified the need to bridge-the-gap between science and engineering and between scientists and engineers. The author echoes this view, because in many instances it has not been feasible to translate or transfer concepts from the different technologies. For example, the author was not able to adopt concepts such as those reviewed by Beaudoin *et al.* (1994) on the influence of pore structure on strength of cement systems.

There are many sophisticated techniques, principally in the realm of powder technology, some involving particle-by-particle sequential packing taking account of such aspects as impact, elasticity and rebound during placement and subsequent movement during vibration, transport and discharge. For example, de Schutter and Taerwe (1993) utilise a random particle method, employing 2-dimensional triangulation for the successive placing of particles in space. Other techniques such as fractal geometry are being applied, e.g. Kennedy (1993) utilises this technique to characterise shapes of aggregate and powder particles. As the prime interest in this present work is the overall end-result rather than detailed process, these elegant and interesting methods have not been pursued.

Some of the models rely on successive combination of component materials or size fractions rather than individual particles. For example, Lees (1970a) proceeds successively from coarsest to finest material while Dewar (1983 etc.) proceeds in the reverse direction.

With regard to modelling of fresh concrete most reported models concern void filling. With the exception of the classic approach of Weymouth reviewed by Powers (1968), the modelling of grading curves reviewed by Popovics (1979) and the work of Hughes (1960), most modelling methods post-date the initial work of Dewar (1983). Most workers, including the author, now place reliance on computerisation to remove the tedium of repetitive complex calculations. Indeed, many of the developments, including that of the author, are impracticable without the computer. It is not surprising therefore that the growth in interest in modelling has had to await the ready availability of computers both for analysis and for making the results usable in practice. Thus, today, relative complexity is much less of a barrier with regard to application, although it may remain a barrier with regard to understanding or persuasion.

Numata (1994) has developed a theory-based modification of the methods of Talbot and Richart and of Weymouth. Roy *et al.* (1993), Palbol (1994) and Goltermann *et al.* (1997) favoured use of the Toufar/Aim (1967) model of dry particle packing or a solid suspension model in association with the Rosin-Rammler size distribution parameters (see section 2.1.2 for an explanation of the R-R distribution). Some authors, e.g. Andersen and Johansen (1989), Roy *et al.* (1993), Goltermann *et al.* (1997) and Palbol (1994), utilise ternary diagrams or packing triangles. Sedran *et al.* (1994) and de Larrard (1995) use a theory of viscosity developed by Mooney (1951). An assessment of the Toufar Aim model and that of de Larrard is provided in Appendix D.

Some workers, notably Powers (1968), Lees (1970a), Dewar (1983) and Loedolff (1985) amongst many others favour voids ratio (i.e. voids per unit solid volume) diagrams because they yield straight line boundary relationships while some, e.g. Goltermann *et al.* (1997) favour solid content (i.e. solids per unit volume of container) diagrams possibly because of their simpler visual relation to bulk density.

Some workers, e.g. Dewar (1983) and Hope and Hewitt (1985), apply a systematic approach to design of concrete by first considering cement paste, then mortar and finally concrete recognising the need for an excess of mortar over that required to just fill the voids.

Some workers favour rodded or vibrated densities for assessing densities or voids contents of component materials, whereas others, including the author, present the case for poured loose bulk densities as being more representative of densities of materials in concrete and less likely to produce anomalous information.

Popovics (1990), in a limited review, was rather despairing of the state of prediction of water demand of concrete. He concluded that prediction was unreliable and valid only over a narrow range; the author's original work, Dewar (1983, 1986) was not included in this review. Larsen (1991) reviewed a number of particle packing concepts but without detailed assessment.

Most workers have been concerned with optimising the design of concrete for maximum density, or minimum water demand, but Palbol (1994) and the author also consider the effects, or even benefits, of moving away from the optimum mixture.

Looking to the future, Frohnsdorff *et al.* (1995) predicts that virtual cement and concrete technology will significantly influence development, through integration of knowledge systems, complementary simulation models and databases. Consistent with part of this prediction, Foo and Akhras (1993) describes a knowledge-based expert system for concrete mixture design, based upon computer processing of input conditions against a substantial accumulated database for the development of a best solution. Andersen and Johansen (1989) reported on the computer-aided simulation of particle packing in relation to the American Strategic Highway Research Program, which is based on the work of Roy *et al.* (1993) who also combined theory with the database approach. Eventually, it is reasonable to expect that, as predicted by Frohnsdorff, a preferred approach or set of approaches will arise adopting the best of the developed systems, but at the present time the field is wide open.

Many workers, e.g. Chmielewski *et al.* (1993), are concerned with comprehensive modelling techniques, but there are a number of specialized design models for special applications, e.g. Dunstan (1983) has developed procedures to produce optimum packing of high fly ash content concretes compacted by vibratory rollers and has also investigated strength; Marchand *et al.* (1997) have reviewed the field of mix design of roller compacted concretes; Mortsell *et al.* (1996) utilised a two phase model, i.e. mortar and coarse aggregate, and developed a new test to assess the mortar fraction; Glavind and Jensen (1996) have commenced exploring the possibility of designing concrete on the basis of packing theory to underfill the voids with paste and provide stable air voids of the required size for frost resistance without recourse to air-entraining admixtures.

There are many very useful and well established concrete mixture design methods, having as their main purpose the recommended proportions for trial concrete batches to be made and adjusted in the laboratory, before transfer to production. Many of these were developed before general computer availability and rely on simplification. For example, most assume that water content is constant over a wide range of cement content.

Unfortunately, these simplified methods are of little use to concrete producers wishing to work with economy for thousands of cubic metres of concrete. For this reason, the prime aim of the author is to develop methods which are more accurate than a single trial in the laboratory with a single set of samples and much more accurate than is possible with simplified approaches.

1.2 The investigation

To achieve this aim, the author has used test data on aggregates to develop a general Theory of Particle Mixtures, based on considerations of void filling and particle interference, involving geometrical and mathematical models and the analysis of voids ratio diagrams. The author is reliant on the concepts of Powers (1968 and earlier) in providing the foundation on which the theory has been constructed. Empirical constants have been used sparingly.

The theory has then been extended to cover cement pastes, mortars and concretes and the inclusion of additional fine powders and plasticising and air entraining admixtures. The properties covered include water demand, slump, cohesion, cement content, plastic density and compressive strength.

This extension has been based on trials made by the author or under the author's direction, on trial data provided by industry and on the published literature. Again empirical constants have been utilised and to a greater degree in the extension of the theory.

Many examples are provided comparing theoretical relationships or inferences with practice and a number of case studies are included to demonstrate the use of the theory as a diagnostic and development tool.

Uncertainties remaining unresolved are identified as recommendations for consideration in any related future work.

The data have been analysed using Microsoft Excel v5.0 spreadsheet software making use where necessary of the graphical and statistical analysis tools and the 'Solver' solution optimization tool developed by Lasdon *et al.* (1978).

2 The principal properties and test methods

The basic concept of the Theory of Particle Mixtures, shown in Figure 2.1, is that, when two particulate materials of different sizes are mixed together, the smaller particles will attempt to fill the voids between the larger particles, but the structure of both coarse and fine materials will be disrupted by particle interference creating additional voids. The modelling process and associated Theory of Particle Mixtures require relevant properties to be defined and appropriate test methods to be selected in order to quantify the effects.

The main relevant properties are mean size, voids ratio and relative density, the latter being required only to convert data from mass to volume. The test methods in Table 2.1 were selected on the basis of perceived relevance to these properties, their common use in industrial practice and the availability of data.

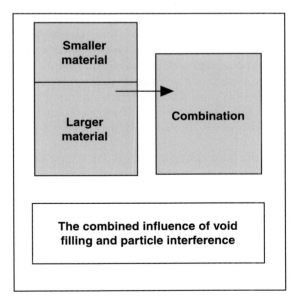

Figure 2.1 The basic principle of the Theory of Particle Mixtures.

Table 2.1 Properties and test methods

Property	Test method	
	For powders	*For aggregates*
Mean size	Particle size distribution	Grading
Voids ratio*	Vicat test for water demand	Bulk density (loose ssd)**
Relative density	Relative density	Relative density (ssd)**

* Ratio of sum of air and water voids to volume of solids.
** Saturated and Surface Dry.

In the event, it was not found necessary to develop new or modified test methods but, in the case of cement mean size, it was necessary to identify alternative methods that could be used in the absence of data based on the preferred method. The details of the main methods are defined in the British Standards for testing aggregates and cements.

2.1 Mean size

2.1.1 *Mean sizes of aggregates*

Popovics (1979) considered average particle size as a more accurate numerical characteristic of the coarseness or fineness of an aggregate fraction than maximum or minimum size. Popovics identified four measurements of average particle size: these are arithmetic, geometric, harmonic and logarithmic averages. Other possibilities include average volume diameter and specific surface diameter, both based upon equivalent spheres.

When dealing with nominally single-sized materials having a relatively narrow size distribution, the definition used has relatively little effect on the value of mean size but for widely graded materials the definition is much more significant.

There is no consensus on the most relevant method. For example: Hughes (1960) used mean size on an arithmetic basis rather than a logarithmic basis; Powers (1968) preferred geometric mean because cumulative plot to semi-log scales tends to be linear; Numata (1995 pc) adopted a logarithmic basis for assessing mean sizes of coarse aggregates but the average volume diameter for fine aggregates; Goltermann *et al.* (1997) and others have adopted the position parameter of the Rosin-Rammler distribution (see Section 2.1.2 for an explanation of the R-R distribution).

Thus, size characterisation is an area where more work could be usefully done to resolve the question of the most relevant method. Pro tem, the author has continued to use the logarithmic basis adopted by Dewar (1983).

Thus, the mean size[1] of a single-sized material is calculated from

$$\text{Log(mean size)} = 0.5[\log(\text{upper size}) - \log(\text{lower size})] \tag{2.1}$$

For a graded material, the mean size is calculated from the grading as

$$\text{Log(mean size)} =$$
$$\sum[\text{vol. propn} \times \log(\text{meansize of the size fraction})] \tag{2.2}$$

The size ratio of two materials to be combined in a mixture is defined as

$$r = \frac{\text{mean size of the smaller material}}{\text{mean size of the larger material}} \tag{2.3}$$

Thus r is always $\leqslant 1$.

Examples of mean sizes and size ratios for materials for concrete are shown in Table 2.2.

2.1.2 *Mean sizes of cements and other powders*

Mean sizes of powders, e.g. cements; fly-ash; ground-granulated blastfurnace slag are determined in the same way as for aggregates, using the particle size distribution, usually on the basis of sedimentation but sometimes on the basis of laser granulometry. It is recognised that results are affected by the method chosen, particularly for the finest sizes and either standardisation or accurate conversion is recommended to be considered for future work.

When the full particle size distribution is not available, it is necessary either to assume the mean size on the basis of experience or to resort to alternative methods.

For cements, fineness measurements are commonly available and estimates of mean size may be made from them. The Blaine values were the commonest basis used. Again it is recognised that other methods produce different answers and that standardisation or accurate conversion is needed for future work.

Table 2.2 Examples of mean sizes and size ratios for materials for concrete

Material	Mean size (mm)	Size ratio r
Cementitious material	0.015	1/40 = 0.025
Fine aggregate	0.6	1/20 = 0.05
Coarse aggregate	12	

Fineness is related simply to the reciprocal of mean size, assuming that shape is constant. UK data suggest that fineness of Portland cement and other powders can be converted approximately to mean size as shown in Figure 2.2, using the formulae in equation 2.4 or equation 2.5.

$$\text{Mean size} = k_f/(RD_p \times F) \text{ mm} \tag{2.4}$$

where

RD_p is the relative density of the powder and
F is the fineness (Blaine) in m^2/kg
k_f is a constant typically 12 to 15 (14.4 from UK data; 12.2 from German data)

or for a Portland cement, assuming a relative density of 3.2 (see later) and using $k_f = 14.4$

$$\text{Mean size} = 4.5/F \text{ mm} \tag{2.5}$$

Cement practitioners and researchers, e.g. Rendchen (1985) in Germany and Sumner (1989) in the UK, commonly transform the particle size distributions (psd) of cements and other ground materials by means of the Rosin-Rammler distribution function which characterises a distribution by a size (or 'position') factor, x_0, and slope, n. The size factor, x_0, is that associated with 63% of the material passing. The mean size on a log basis d_p used by the author corresponds to about 50% passing and is thus always lower than x_0. For example a typical cement might have a mean size d_p of 13 μm compared with a value of 22 μm for x_0.

When the value of x_0 is known, but the full psd is not available, then it can be assumed approximately that

$$d_p = 0.64 \times x_0 \times n \quad \mu m \tag{2.6}$$

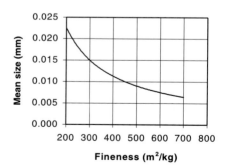

Figure 2.2 Approximate relationship between mean size and fineness (Blaine) for Portland cement (UK data).

or more accurately from

$$d_p = 1.06 \times e^{(\log_n(x_0) - 0.53/n)} \qquad \mu\text{m} \qquad (2.7)$$

Rendchen (1985) has published data on 22 cements, having a wide range of properties, for which a summary is provided in Table 2.3 together with the results of an analysis made by the author to examine different means of assessing mean size from the data.[2]

The methods investigated for estimating mean size were

1. Calculation direct from the reported particle size distribution.
2. The formula derived from the fineness and relative density (equation 2.4).

Table 2.3 Data on 22 cements as reported by Rendchen (1985) together with an analysis by the author of different methods of assessing mean size on a log basis

Cement code	Fineness m^2/kg	R–R constants		Mean size(log basis) (mm)				
		x_0	n	From psd	From fineness	From R-R distn	Accurate from x_0 & n	Approx from x_0 & n
H1	271	31	0.85	0.0160	0.0144	0.0173	0.0176	0.0168
H2	293	32	0.82	0.0169	0.0135	0.0175	0.0178	0.0167
H3	300	28	0.89	0.0153	0.0132	0.0160	0.0164	0.0158
H4	308	28	0.79	0.0141	0.0126	0.0152	0.0152	0.0141
H5	321	27	0.87	0.0147	0.0122	0.0154	0.0156	0.0149
H6	282	18	1.09	0.0117	0.0138	0.0113	0.0117	0.0125
H7	344	19	0.91	0.0108	0.0114	0.0111	0.0113	0.0110
H8	361	14	1.11	0.0103	0.0109	0.0090	0.0092	0.0099
H9	373	16	0.94	0.0097	0.0107	0.0095	0.0097	0.0096
H10	371	13	1.04	0.0089	0.0107	0.0082	0.0083	0.0086
H11	535	11	0.93	0.0069	0.0073	0.0067	0.0066	0.0065
H12	374	14	1.02	0.0090	0.0110	0.0087	0.0088	0.0091
NH1	403	17	0.95	0.0104	0.0098	0.0102	0.0103	0.0103
NH2	487	18	0.78	0.0095	0.0081	0.0100	0.0097	0.0089
NH3	602	11	0.89	0.0065	0.0066	0.0066	0.0064	0.0062
NH4	437	13	1.06	0.0093	0.0091	0.0082	0.0084	0.0088
NH5	517	11	1.08	0.0073	0.0077	0.0070	0.0071	0.0076
NH6	418	14	1.07	0.0098	0.0094	0.0088	0.0090	0.0095
S1	347	18	0.92	0.0106	0.0112	0.0106	0.0107	0.0105
S2	523	13	0.86	0.0078	0.0076	0.0076	0.0074	0.0071
S3	341	14	1.15	0.0097	0.0113	0.0091	0.0094	0.0102
S4	510	10	1.03	0.0066	0.0076	0.0064	0.0063	0.0066
			Average	0.0105	0.0105	0.0105	0.0106	0.0105

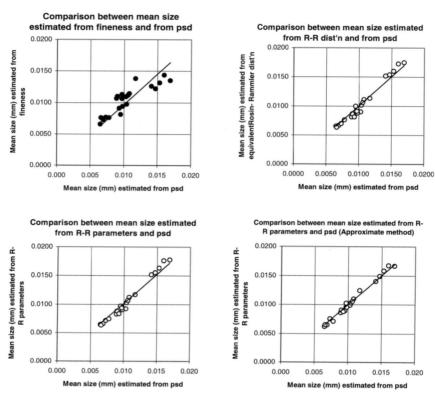

Figure 2.3 Comparisons of methods of assessing mean size of cement, the basis of the comparison being the mean size on a log basis calculated from the reported particle size distribution.

3.   Calculated straight-line distributions estimated from the reported Rosin-Rammler parameters.
4.   Calculation from the Rosin-Rammler parameters assuming that the mean size was coincident with the 50% point of the size distribution (equation 2.7).
5.   Calculation using a simplified formula involving the Rosin-Rammler parameters (equation 2.6).

Taking Method 1, calculation direct from the psd, as the basis for comparison, the validity of the various other methods of assessing mean size can be judged from examination of Table 2.3 and Figure 2.3.

As might be expected, Method 2 utilising Blaine fineness is the least accurate, but is nevertheless the most important method for concrete, because of the relatively greater availability of data on fineness in everyday practice, compared with data on particle size distribution. In addition, it may be noted again that the different methods of measuring the particle size distribution have their own problems of interpretation, as do the various fineness methods.

For some very fine materials, e.g. silica fume, the mean particle size may be uncertain, because of agglomeration and may be effectively much greater than suggested by the literature. This is considered further for a particular type of very fine powder in a case study in Section 8.1.

2.2 Relative density

2.2.1 *Relative densities of aggregates and fillers*

It is assumed that the relevant volume of an aggregate particle is normally the overall particle volume, including the pores occupied by air or water. The relevant relative density for calculating this volume from tests of mass in air, e.g. bulk density tests, will be

- Relative density on a saturated and surface dried basis to BS 812, if the bulk density tests were made on saturated and surface dried aggregates.
- Alternatively, relative density on an oven-dried basis to BS 812, may be used if the bulk density tests were made on oven dried aggregates.

It is assumed that the aggregates are used generally in production of concrete, in both the plant and laboratory, in a saturated condition, either as delivered and stored or after saturation, if pre-dried for laboratory purposes. Materials in other conditions, e.g. dry lightweight aggregate used dry, may need special consideration to ensure that the principles and calculations in this book remain valid. Fully saturated lightweight aggregates are fully within the scope.

For concrete calculations, the relevant relative density value of aggregates for converting from volume to mass is the

- Relative density on a saturated and surface dried basis.

This is compatible with the absorbed water and residual internal air being considered part of the volume and mass of the aggregate; the void content of the concrete is thus the sum of the volumes of free water and air external to the aggregate.

2.2.2 *Relative densities of cements and other powders*

Powers (1968) identified that the relative density of Portland cement in water was about 0.06 greater than in kerosene due to the solubility of the calcium sulphate and other components and that the difference varied between cements.

As cement operates in an aqueous environment in concrete, it is appropriate to use the value for relative density of Portland cement in water which is taken to be 3.2, corresponding to 3.12–3.15 in kerosene.[3]

For additions, e.g. fly ash, the relative density may be measured either in water or in kerosene.

For cements containing additions, allowance needs to be made for the relative densities of the components and for their proportions.

Typical values for *relative densities* of cementitious materials are

- Portland cement   3.2 (equivalent to 3.12–3.15 in kerosene)
- Cements containing additions or fillers   Value depends on proportions of components
- Ground granulated blastfurnace slag   2.9
- Fly-ash   2.3
- Silica fume   2.2
- Fine limestone filler   2.7

2.3 Voids ratio

The voids ratio of a particulate material is defined as the ratio of voids to solids under a stated method and energy level of compaction.

Factors, such as the range in particle size about the mean size and the shape and texture of particles, are accounted for in the assessment of the voids ratio of a material.

2.3.1 Influences on voids ratio

It is necessary to consider the medium in which the voids ratio is to be measured, i.e. whether the tests should be made in air or in water, and the energy level to be adopted in placing and compaction.

Influence of moisture environment on compaction

It is well known that the bulk densities of particulate materials, more particularly powders, are influenced markedly by the moisture condition at test. The phenomenon of bulking of sands and powders at low intermediate moisture contents compared with dry or inundated conditions is to be avoided when simulating conditions applying in most normal concretes. Equally to be avoided is the bulking of dry powders.

Some workers, notably Powers (1968) and Loedolff (1985) have investigated the voids ratios of dry powders and dry mixtures of sand and cement under high energy compaction. Loedolff reported anomalous results on occasion in comparison with wet mixtures. This is not surprising for two reasons, viz.: the massive bulking effects with dry powders and the effects described in Appendix A due to segregation under high energy compaction.

For typical aggregates, the voids ratios of the resulting mixtures are such that they may be simulated by compaction in air or water, but air is favoured for convenience by most workers including the author.[4]

For powders, the interparticle forces in air are substantially higher than in a saturated environment such as fresh concrete. Thus, it is more relevant for powders to be tested in a water medium at a consistence corresponding to typical concrete, as in the Vicat test.

In support of this, a number of workers have either adopted the Vicat test or have developed similar tests. For example, Chmielewski *et al.* (1993) adopted the Vicat test as a measure of water demand of cement; Puntke (1996) developed a method for assessing fine components separately or together by determining voidage at saturation in water; de Larrard and Sedran (1994) measured packing density of cement as a water–cement paste incorporating a plasticiser as necessary. The method of judging consistency was not described by de Larrard.

Contributions by Murphy and Newman to discussion of Murdock (1960) confirm the significance for concrete production of differences in the water demand of cement in the Vicat test.

As a further justification for using the Vicat test, Powers (1968) reported that typically, cement paste of normal consistency yields a slump of about 40–50 mm. This was confirmed in laboratory tests by the author.

Influence of the energy level of compaction

Opinion is divided on the question of the compactive effort to be adopted when simulating conditions of particles in most conventional concretes. It is clear that some workers have been attracted to higher energy levels, e.g. rodding or vibration, on the grounds of better reproducibility of test results and because of the approach to a lower limiting value for voids ratio or because of particular construction situations under consideration.

All methods of placing in air, with or without compactive effort, rely on displacement of the air and in the process may encourage segregation of the placed material, which in turn will introduce variations in voids ratio locally within the mixture and affect the overall value.[5]

There are at least three main types of segregation to be avoided

- Flow of fine particles downwards between coarse particles under the action of gravity, rodding or vibration
- Movement of coarse particles downwards and fine particles upwards under the action of rodding or vibration
- Movement of coarse particles upwards due to a circulatory or convection effect induced by strong vibration and the container walls

With loose poured packing only, the first is likely to be significant and only for low proportions of fine material. Rodding or vibration may introduce the other types of segregation.

Considering first those workers favouring loosely compacted methods

- Hughes (1960) developed various parameters for size and voidage of aggregates. In particular, Hughes favoured loose bulk density as more relevant than vibrated bulk density because the particles are 'relatively free to reorientate themselves under any external influence'.
- Bloem and Gaynor (1963) considered that loose void content of fine aggregate was a useful means for evaluating effect on water demand of concrete. Bloem also considered that fine aggregate particle shape and texture had a readily discernible effect on water demand of concrete.
- Hughes (1976) favoured loose bulk density for assessment of effect of coarse aggregate on water demand of concrete.
- Gaynor and Meininger (1983) favoured a loose voids test for assessing sands for concrete with respect to effects of shape and texture.
- Johnston (1990) considered that the loosely packed (poured) condition reasonably represents the aggregate packing in concrete while it is being placed before consolidation.
- Brown (1993) favoured loose bulk density of aggregates as a means of assessing the combined effects of grading and shape.
- Numata (1995) observed that the state of fine aggregate in concrete corresponds to the loosely packed state because of the presence of powder.

Considering next those workers having divided or intermediate views

- Wills (1967) based assessments of water demand of concrete on loose void content of fine aggregate and the dense value for coarse aggregate,
- Powers (1968) favoured rodded, rather than loose or vibrated, bulk density while recognising that voids measurement is more sensitive to differences in shape, if it is based on the loose aggregate volume.
- Goltermann *et al.* (1997) adopted a hand operated jolting table.

Considering those workers who favour higher energy compaction

- Talbot and Richart (1923) utilised rodded bulk density (b_0) test of coarse aggregates for assessing the coarse aggregate content (b) of concrete by application of a reduction factor (b/b_0) of 0.65–0.75, typically. Talbot recognised that (b/b_0) could not reach unity because of wedging action by the mortar and that coarse aggregate particles in concrete might not arrange themselves as well as when measured alone. The (b/b_0) concept continues to be used in the ACI method of concrete mix design today.

- Stewart (1962) considered it to be important for the method of compaction of aggregates to be comparable with that used in concrete. Being concerned more with vibrated concretes, Stewart adopted vibration for assessing aggregates.
- Bloem and Gaynor (1963) used a factored volume assessed from the dry-rodded bulk density of coarse aggregate to calculate the volume to be used in all concretes irrespective of cement content. The multiplying factor varies between 0.35 and 0.85 dependent upon the fineness of the sand and maximum size of coarse aggregate.
- Loedolff (1986) determined packing curves under intense vibration exceeding 20 mins for dry and wet (75 mm slump) materials, for individual materials and in combinations including powders.
- Roy (1993) utilised rodded bulk densities of the aggregates and vibrated bulk densities of dry powders.
- The ACI (1993) uses the volumes of oven-dry rodded coarse and fine aggregate as a means of assessing the quantities required for concrete. For coarse aggregate this requires a substantial reduction factor, e.g. 0.72, to be applied to obtain the quantity to be used in practice to allow for interference from the sand and to allow for normal workability.

Finally, considering those workers who have compared methods

- Moncrieff (1953) compared bulk densities of natural and crushed fine aggregates in the loose and rodded condition. Whereas compacted natural sand had a bulk density about 5% higher than in the loose condition, the difference increased to about 10% for the crushed stone. Thus, two materials having the same bulk density and voids ratio in the loose state could be expected to differ by about 5% in bulk density when compared in the compacted state, with the crushed material yielding the higher value. Thus, it is important firstly to decide on the most appropriate method to simulate practice and then to use the chosen method consistently.
- Hughes (1962) concluded that measurement of maximum bulk density under vibration was unreliable due to effects of containers and due to difficulty in determining a suitable end point. It was possible for continued vibration to decrease bulk density. Hughes adopted minimum or 'loose' bulk density because it was easier to determine, confirming an earlier view of Worthington (1953).

Unfortunately, there is not a simple conversion from rodded or vibrated density to, or from, loose bulk density. Indeed an angular material may have a lower voids ratio than a round material under vibration but a higher voids ratio under loose packing conditions. Thus, the conversion process is suspect and is better avoided.

Some techniques may be more likely to promote forms of segregation which may not occur, or be less likely to occur, with concrete and with a low energy method. Thus, on balance, it is preferable to use a test that may be more relevant and sensitive, at the possible expense of precision and one that is less likely to introduce significant effects of segregation. For concretes of medium and high workability in common use today, the author considers that the loose bulk density test in air is more relevant for both fine and coarse aggregates.

Of course, there may be construction situations when a rodded or vibrated condition may be more appropriate than a loose condition, e.g. concrete of very low workability to be compacted by heavy vibration and/or pressure. These were not considered to be within the scope of the author's research.

Further considerations of the effects and pitfalls of high energy compaction of aggregates are provided in Appendix A.

2.3.2 Voids ratios of aggregates

Taking account of the discussion on moisture condition and energy level, the BS 812 method of assessing the loose bulk density of aggregates was selected as appropriate for work relating to mortars and concretes but utilising the smaller container (approx. 0.003 litre) for both fine and coarse aggregates.

The voids ratio of an aggregate can be calculated from its bulk density in air and its relative density.

The formula for calculating the voids ratio, U, of an aggregate is

$$U = \frac{1000 \times RD}{LBD} - 1 \tag{2.8}$$

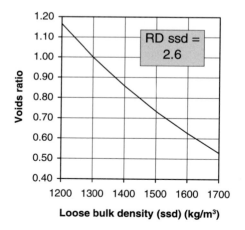

Figure 2.4 Example of a relationship between voids ratio and loose bulk density of aggregate (ssd).

where *RD* is the relative density of the aggregate and *LBD* is the loose bulk density of the aggregate.[6]

An example of a relationship between voids ratio and loose bulk density of aggregate is shown in Figure 2.4.

2.3.3 Voids ratios of cements and other powders

The voids ratio *U* for a powder can be estimated from the Vicat test using the following formula

$$U = (RD_p \times SC + a_p)/(100 - a_p) \qquad (2.9)$$

where

RD_p	is the relative density of the powder
SC	is the water content of the paste at standard consistence in the Vicat test to EN 196, calculated as a percentage of the mass of cement
a_p	is the air content (%) in the paste, say 1.5.

For Portland cements, a typical relation between voids ratio and water demand of the cement paste is shown in Figure 2.5.

A voids ratio of 0.825 corresponds to a typical paste water demand of 25%. Water demands of over 30% correspond to voids ratios of over 1, implying that the paste contains more voids than solids.

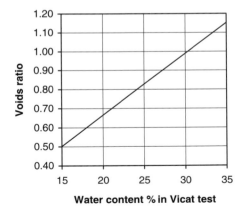

Figure 2.5 Relationship between voids ratio and water content on the Vicat test for Portland cement having a relative density of 3.2 in water.

Table 2.4 Typical values for the main properties of cements and aggregates based on definitions and test methods adopted by the author

Material	Mean size (mm)	Voids ratio	Relative density
Cement	0.013	0.83	3.20
Fine aggregate	0.50	0.70	2.60
Coarse aggregate	11	0.80	2.55

2.4 Overview concerning properties of materials and test methods

Three properties only are required to characterise each material for prediction of properties of fresh mortars and concretes. These properties are mean size, voids ratio and relative density. They can be estimated from commonly available test data without need for additional tests or special equipment.

Examples of values for common concreting materials are shown in Table 2.4 based on the definitions and test methods adopted by the author.

Application of measurement of size and fineness of powders could benefit from rationalisation of methods or publication of conversion factors, to minimise difficulties of interpretation of data from different sources based on different methods.

Characterisation of 'mean' particle size could also benefit from rationalisation.

The adoption of loose bulk density (for assessing voids ratios) of particulate materials is recommended rather than rodded or vibrated densities to minimise segregation. It is more relevant for bulk densities (for voids ratios) of powders and possibly also fine sands, fine fractions of sands and fillers to be determined in water rather than air to minimise effects of agglomeration.

3   Theory of Particle Mixtures

In this section, the Theory of Particle Mixtures is developed from analysis of tests of aggregates firstly to cover mixtures of

- Two single-sized components
- Two graded components
- Three or more components

before extension to cover mortars and concretes in section 4.

3.1 General theory of particle mixtures developed for mixtures of two single sized components

Visual and mathematical models are used to explain the theory and to show how the various constants and formulae have been derived. In particular, the value of the voids ratio diagram is emphasised for understanding the interactions between the various mechanisms.

Some comparisons between theoretical and actual voids ratio diagrams are provided to assist judgement and to anticipate the more comprehensive comparisons in later sections, without distracting from the general flow of the development.

It is assumed that significant segregation does not occur.

3.1.1 Modelling of particles and voids

For analytical purposes, each real particle in a mixture of single-sized particles is assumed to be associated with a single corresponding void, as in Figure 3.1. In reality, of course, each solid particle shares voids with other particles and the voids are effectively continuous, but the net effects can be modelled as an equal number of finite voids and particles. Thus, the combination of a single particle and its associated 'single' void can be considered to represent a key 3-dimensional composite element in the mixture of particles.

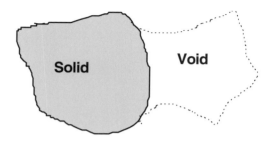

Figure 3.1 Particle and associated void.

The **voids ratio** for the particular material, as used by Powers (1968) and in this book, is defined as the ratio of total void volume to total solid volume determined by compaction under stated conditions (section 2.3).[1]

The voids ratio is a function of the shape and surface texture of particles and of the grading about the 'mean' size of the particles.

For analysing voids ratio diagrams, the author has developed a number of 3-dimensional geometric models, such as those shown in Figure 3.2 and Figure 3.3.

For both analytical purposes and for easy comprehension, a particle of volume D^3 is modelled as a cube having a 'mean' size D. It should be understood that a cube is used merely for convenience in the visual model and there is no assumption that this is a real, common or ideal shape.

By adopting a convenient arrangement for the composite structure as in Figure 3.2, it is possible to model any mean size and voids ratio for any given particulate material. The use of an expanding or contracting 'iris' or camera

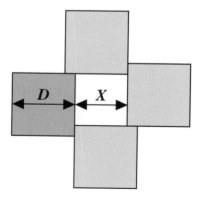

Figure 3.2 Three-dimensional model of a particle, associated void and related particles.

diaphragm structure enables any voids ratio between 0 and unity to be accommodated by increasing or decreasing X, the side length of the cubical void associated with a particle.

For the case, when $X \leqslant D$, then each particle is associated with a corresponding void volume as in Figure 3.2 where X is the mean size of void and D is the mean size of particle.

The voids ratio is given by

$$U = \frac{X^3}{D^3} \tag{3.1}$$

Thus

$$X = D\sqrt[3]{U} \tag{3.2}$$

For the particular case when the voids ratio is unity, the coarse particles are in contact only at the corners.

Where the nature of the material results in a voids ratio greater than unity, e.g. due to a very unfavourable real shape, or due to agglomeration of particles, then the model needs to be considered to expand beyond $X = D$ to give notional 'cubical' particles not in contact.

When differentiating between different components used in a two-size particle mixture, a suffix 0 is attached to parameters for the larger (coarse) component and 1 for the smaller (fine) component. These suffixes 0 and 1 are also, respectively, the volumetric proportions of the fine component in a two-size particle mixture for the two extreme situations of all coarse and all fine material.

Thus, for the coarse particles alone

$$U_0 = \frac{X_0^3}{D_0^3} \quad \text{and} \quad X_0 = D_0\sqrt[3]{U_0} \tag{3.3}$$

and for the fine particles alone

$$U_1 = \frac{X_1^3}{D_1^3} \quad \text{and} \quad X_1 = D_1\sqrt[3]{U_1} \tag{3.4}$$

The mixing of fine particles and coarse particles dilates the structure of the coarse particles. This is modelled by assuming that the coarse particles move apart, as in Figure 3.3, to occupy the centres of spaces which have the same geometric relationship with the contained void, as existed in the model before dilation, i.e.

$$\frac{X_0''}{(D_0 + mD_1)} = \frac{X_0}{D_0} = \sqrt[3]{U_0} \tag{3.5}$$

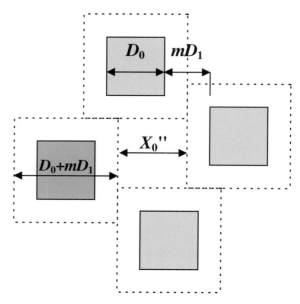

Figure 3.3 Three-dimensional model of the dilated structure of coarse particles in a
mixture containing fine particles.

where the coarse particles are spaced apart m times the fine particle mean size
D_1.

The effective voids ratio of the coarse particles when the structure is dilated
is

$$U_0'' = \frac{(D_0 + mD_1)^3 + (X_0'')^3 - D_0^3}{D_0^3} = (1 + mr)^3 + \frac{(X_0'')^3}{D_0^3} - 1 \qquad (3.6)$$

where the ratio of mean sizes is

$$r = \frac{D_1}{D_0} \qquad (3.7)$$

but from equations 3.5 and 3.7

$$(X_0'')^3 = U_0(D_0 + mD_1)^3 = U_0(1 + mr)^3 D_0^3$$

i.e.

$$(X_0'')^3 / D_0^3 = U_0(1 + mr)^3$$

Substituting in equation 3.6 then

$$U_0'' = (1 + mr)^3 + U_0(1 + mr)^3 - 1$$

which simplifies to

$$U_0'' = (1 + U_0)(1 + mr)^3 - 1 \qquad (3.8)$$

If U_0'' and U_0 are known then the spacing factor m can be estimated from equation 3.8 by transposition

$$m = \frac{1}{r} \left[\sqrt[3]{\left(\frac{1 + U_0''}{1 + U_0}\right)} - 1 \right] \qquad (3.9)$$

In section 3.1.2 a method is described for estimating U_0'' and m graphically using a voids ratio diagram.

3.1.2 Voids ratio diagrams

Voids ratio diagrams, introduced by Powers (1968 and earlier) are essential tools for understanding the processes involved in the combination of particulate materials. Figures 3.4–3.7 show the basic elements of such diagrams.[2]

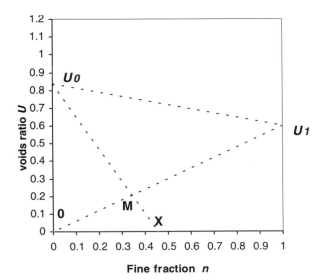

Figure 3.4 Example of a voids ratio diagram.

These types of diagram have been adopted by Lees (1970) and Al Jarallah and Tons (1981) for dense asphaltic compositions, by Dewar (1983) and Loedolff (1986) for aggregates and concretes and by Kantha Rao and Krishnamoothy (1993) for polymer concretes. Lees for example, identified size ratio, particle shape and texture, void ratios of components, container size and compactive effort as relevant parameters determining the shape and position of the diagram. Some workers, while utilising the concept, have used it primarily for display purposes and others, notably Powers (1968), have used it for analytical purposes and as a design aid. Loedolff (1986) utilised laboratory obtained void ratio diagrams of aggregates before introducing finer materials e.g. cement, fly-ash, ground slag or silica fume assessed in the same way. Dewar (1983) used the diagrams to develop the first version of a Theory of Particle Mixtures and applied it to aggregates, mortars and concretes.

To produce a voids ratio diagram, the voids ratios of the coarse and fine components, U_0 and U_1, of a two-material mixture are plotted on the left and right vertical axes respectively. The fine fraction on a volumetric basis, n, is plotted on the horizontal axis.[3]

The line U_0X represents the theoretical effect on voids ratio of adding fine material without any dilation of the structure of the coarse material. Such a situation is uncommon and would apply only if the coarse material was held rigidly in place while fine material was poured or vibrated through the structure. M is the point at which the voids in the coarse material are just filled with the fine material at its own voids ratio while the coarse particles remain in contact. If the line U_0X is continued it will always cross the right-hand axis at $n = 1$ and $U = -1$.[4]

The line $U_{1,0}$ represents the theoretical effect of addition of coarse material to fine material without increasing the voids ratio of the fine material. Such a situation would require the coarse material to displace both fine material and associated voids without changing the structure of the unreplaced fine material. M is the point at which the coarse particles come into contact.

M_0 and MX represent hypothetical mixtures which cannot exist.

The lower boundary, U_0MU_1, represents the voids ratios for all combinations of the two materials that could exist, but only in the absence of particle interference. Under certain conditions, real mixtures may approach the boundaries.

The triangular area U_0MU_1 represents all practical mixtures that can exist in the presence of particle interference.

The upper boundary, U_0U_1, represents the case when the mean sizes of the components are equal and interference is thus at a maximum. Puntke (1996) confirmed that when materials of similar grading were combined, the resulting voidages of combinations are related linearly to the proportions of the materials.

This boundary diagram can be drawn simply by joining, with straight lines, points U_0 and U_1 and points with co-ordinates $(0, U_0)$ and $(1, -1)$ and the points $(0, 0)$ and $(1, U_1)$.

If point $(1, -1)$ is inconvenient, point X can be substituted using equation 3.14.

The voids ratio in the range U_0M in Figure 3.4 is

$$U_n = U_0 - n(1 + U_0) \tag{3.10}$$

The voids ratio in the range MU_1 is

$$U_n = nU_1 \tag{3.11}$$

From equations 3.10 and 3.11 the co-ordinates of M can be calculated as

$$n_M = U_0/(1 + U_0 + U_1) \tag{3.12}$$
$$U_M = U_0U_1/(1 + U_0 + U_1) \tag{3.13}$$

In the example in Figure 3.4, substituting the values for U_0 and U_1 of 0.835 and 0.6 in equations 3.12 and 3.13 yields

$$n_M = 0.345 \quad \text{and} \quad U_M = 0.205$$

Thus, by simple theory and in the absence of particle interference, the voids ratio could be substantially reduced from 0.835 to 0.205 by combining the fine and coarse materials in the proportions by volume, 34.5% : 65.5%.

In order to draw the left-hand boundary in the voids ratio diagram without plotting $(1, -1)$, point X can be substituted. From equation 3.10 the fine fraction for point X can be calculated as

$$n_x = U_0/(1 + U_0) \tag{3.14}$$

In the example in Figure 3.4 for $U_0 = 0.835$, $n_X = 0.455$.

It will be apparent that, assuming no particle interference, a lower voids ratio for either or both of the components yields lower boundaries and a lower minimum voids ratio at point M.

Figure 3.5 shows a comparison between experimental data and the theoretical relationships from Figure 3.4.

It will be seen that the experimental data lie above the lower theoretical boundary, generally within the triangle as postulated, and that the deviation vertically from the lower boundary is greatest in the vicinity of M.[5]

The proximity or otherwise of the experimental values to the lower boundary is a function of the size ratio, r. A relatively low value of r, e.g. 0.059, as in the example in Figure 3.5, implies low particle interference. A high value for r, approaching unity, would lead to results near the upper boundary. However, even with a value of r as low as 0.059, the lowest voids ratio achieved of 0.33 was over 50% higher than predicted at M by simple theory. Thus, particle interference can be seen to have a major influence on the voidage of mixtures and should not be ignored. Indeed, any theory that

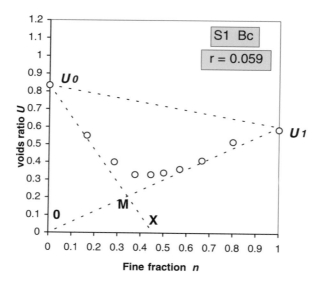

Figure 3.5 Comparison of experimental data and theoretical relationships for voids ratio.

does not take account of particle interference is likely to be wholly inadequate in the vicinity of *M* unless the size ratio *r* is extremely low.

Overview of simple theory in relation to voids ratio diagrams

Basic theory of void filling in the absence of particle interference coupled with voids ratio diagrams provide simple visual means for analysis and judging the effects of particle interference on real mixtures.

3.1.3 Analysis of voids ratio diagrams for real particle mixtures

Worthington (1953) recognised that a simple theory of packing as described in section 3.1.2 was inadequate, due to particle interference. Powers (1968) observed that when fine aggregate is mixed with coarse aggregate, fine aggregate not only accommodates itself in the voids but disperses the coarse particles. Powers distinguished fine aggregate dominant mixtures from those which were coarse aggregate dominant and recognised that particle interference provided the reason for major departures of actual voids ratio diagrams from the boundary lines for coarse particle dominant mixtures. Powers did not associate the lesser departure for fine particle dominant mixtures with particle interference, whereas the author attributes all departures to be the effect of forms of particle interference.

With voids ratio diagrams and other similar diagrams concerning voids or density, it is natural and commonplace in the literature to find that when relatively few experimental points are plotted, they are represented by a smooth curve. However, if numerous closely spaced and careful experiments are done, e.g. in the work of Powers (1968) and Loedolff (1986), it will be found that the results tend to lie close to a set of straight lines and that some of these lines demonstrate a marked change in slope. It is a contention of the author that there are several change points and each is associated with a particular value for m, the spacing factor.[6]

Change point C corresponds to a particular value of spacing factor, m, of approximately 0.75. Thus at this point the coarse particles are spaced apart $0.75 \times$ the diameter of the fine particles. This can be confirmed experimentally by reading off from Figure 3.6 the value of U_0'' of about 1.26, which is the effective voids ratio at C for the coarse material, and substituting this value in equation 3.9. This yields a value of 0. 745 for m.

Conversely, if the value of $m = 0.75$ is assumed to apply at point C then the voids ratio of the dilated coarse material can be found to be 1.26 from equation 3.8 enabling the position of point C to be identified within the data by joining $U_0'' = 1.26$ to the point $(1, -1)$ below the diagram, as shown by the upper dotted line. The divergence between the two dotted lines in Figure 3.6 represents the effects of particle interference and increases with increasing r.

Three further change points B, D and E at spacing factors of approximately 0.3, 3 and 7.5 have been identified during the analysis of 48 diagrams, for a

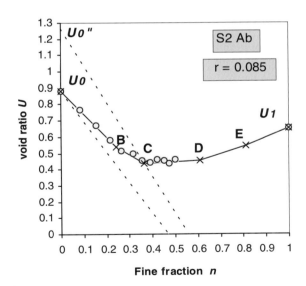

Figure 3.6 Illustration of the effect of dilation of the coarse aggregate structure at change point C, where the spacing factor m is 0.75, causing the effective voids ratio of the coarse material to increase from U_0 to U_0''.

wide range of combinations of single sized and graded materials used for concrete. This analysis confirmed that the value of m was unaffected by voids ratios of the components of the mixture, their size ratio or the energy level of compaction. This is supported by Bache (1981) who considered that the packing of particle systems in which surface forces are insignificant is dependent on the relative size distribution and not on the absolute particle size.[7]

Values of spacing factors for all change points are summarised in Table 3.1.

Just as the effective voids ratio of the coarse material at point C is increased, so also the effective voids ratio of the fine material is increased at the same point by particle interference, as shown in Figure 3.7. Thus, the voids ratio of the fine material effectively increases from U_1 to U_1'' in the mixture.

The fine particles in a mixture occupy the available space within the dilated structure of the coarse particles, but due to particle interference require additional space that may be modelled as a notional volume $0.5 \times Z \times D_0$ wide surrounding the coarse particles as in Figure 3.8.[8]

It should not be assumed that this space exists as a single identifiable entity but as the summation of the increased space required by the effects of particle interference in disturbing the structure of the fine particles, for some distance from the coarse particles. Although most of this extra space will exist close to the coarse particles, some fine particles will be present in this space and some will touch coarse particles.

Powers (1968) accepted an explanation of Weymouth that there was an analogy between boundary effects of container walls and effects of fine particles packing against the surface of coarse particles. However, Powers preferred an alternative explanation that each coarse particle produced a local disturbance reaching beyond the region of direct contact. Bache (1981) also identified two reasons for the density of mixtures not achieving the theoretical

Table 3.1 Summary of values for spacing factors, m, for coarse particles in a mixture with fine particles

Point in voids ratio diagram	Spacing factor, m
A $(n = 0)$	0
B	0.3
C	0.75
D	3
E	7.5
F $(n = 1)$	∞

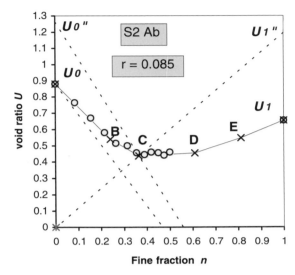

Figure 3.7 Example of the increase of the voids ratios of both the fine and coarse aggregates in the mixture at point C compared with the values for the separate materials, due to particle interference.

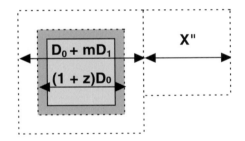

Figure 3.8 Additional voids (indicated by heavier shading) required by the fine aggregate in the proximity of coarse aggregate.

maximum. Bache termed these the 'wall and barrier' effects, where the first relates to the increase in voids at the boundary between coarse and fine particles and the second concerns the lack of space for fine particles between closely spaced coarse particles. Roy *et al.* (1993), also considered that there is a distinct similarity between the boundary effect and the well-documented wall effect associated with particulate materials in a container.

Gray (1968) considered that voidage in a container is affected up to five particle diameters away from a container wall. Powers (1968) observed that when interference commenced, the clearance between coarse particles was about 9 times the average size of the fine particles. The author's proposed value of $m = 7.5$ at change point E, is intermediate between these two values.

As a result of these various effects, the fine particles are at a reduced packing density and thus increased voids ratio compared with the measured value for fine particles alone.

Detailed analyses of data for mixtures of aggregates suggest that the factor Z is a function of the size ratio, r, and the voids ratio U_0 of the coarser material. Thus, the notional width factor of the additional voids is

$$Z = k_{int} + [(1 + U_0)^{1/3} - 1 - k_{int}]r^{k_p} \tag{3.15}$$

where k_{int} and k_p are empirical constants, the values of which depend upon the change point B, C, D or E under consideration. For the example in Figure 3.7 for point C, the values for k_{int} and k_p are respectively 0.06 and 0.65. Substituting these values in equation 3.15 together with the values for r and U_0 of 0.085 and 0.88 respectively, leads to a value of Z of 0.095.

This implies that, in the example, the hypothetical void band surrounding the coarse particle, is $0.5 \times 0.095 \times D_0$, i.e. about $0.05 D_0$ wide. The mean size of the fine material is $r \times D_0$, in this example, $0.06 D_0$. Thus a substantial quantity of space, having a notional width close to that of the fine material mean size, is lost to the fine material due to particle interference under the conditions applying at point C.

An analysis of the geometric and algebraic relationships of Figure 3.7 and Figure 3.8 and the preceding formulae, yielded a relationship by which Z may be estimated from the experimental data, as follows

$$Z = \sqrt[3]{(1 + U_0'') - \frac{(1 + U_1)U_0''}{(1 + U_1'')} - 1} \tag{3.16}$$

where U_0 and U_1 are the voids ratios of the fine and coarse materials respectively and U_0'' and U_1'' are the apparent values of the materials for the point on the voids ratio diagram under consideration. Table 3.2 summarises the results of the analysis.

Table 3.2 Summary of values of empirical factors, for determining Z, the notional width factor of voids at each change point

Point in voids ratio diagram	k_{int}	k_p
B		
C	0.12	0.60
D	0.06	0.65
E	0.015	0.8
	0	0.9

For the example shown in Figure 3.7 for which $U_1 = 0.65$, $U_0'' = 1.26$ and U_1'' is 1.2 then Z for Point C is 0.096 from equation 3.16 close to that of 0.095 determined from equation 3.15.

The apparent voids ratio U_1'' at a point in the voids ratio diagram can be estimated from the following formula, which is a transposition of equation 3.16

$$U_1'' = \frac{(1 + U_1)U_0''}{(1 + U_0'') - (1 + Z)^3} - 1 \tag{3.17}$$

In the same example in Figure 3.7 for point C and taking $Z = 0.095$ then U_1'' is 1.20, as before.

The co-ordinates of the change point can also be estimated as

$$n = \frac{U_0''}{(1 + U_1'' + U_0'')} \tag{3.18}$$

and

$$U_n = nU_1'' \tag{3.19}$$

In the example shown in Figure 3.7, where $U_0'' = 1.26$ and $U_1'' = 1.20$, then the co-ordinates of point C are $n = 0.365$ and $U_n = 0.44$.

In the same way, constants, formulae and co-ordinates for points B, D and E may be determined.

Thus, the voids ratio diagram can be constructed from the co-ordinates of each change point $A-F$ and the quantities of materials per unit volume can also be calculated.

For example, if the voids ratio at C is $U_n = 0.44$, then the voids content V_n at point C is

$$V_n = \frac{U_n}{1 + U_n} \tag{3.20}$$

Thus, V_n is 0.305 m³/m³ and S_n the solids content may be calculated from either

$$S_n = 1 - V_n \quad \text{or} \quad S_n = \frac{1}{1 + U_n} \tag{3.21}$$

Thus, the solids content at C is 0.695 m³/m³.

If the proportion of fine to total solid material is $n = 0.365$ by volume, then the solids content can be proportioned correspondingly as

$n \times S_n$ of fine material

and

$(1 - n) \times S_n$ of coarse material

and can be converted to mass using the appropriate relative densities.

3.1.4 Overview: summary of the key formulae for the construction of voids ratio diagrams and calculations of quantities for mixtures of two components

The effects of mixing particulate materials can be illustrated by voids ratio diagrams such as the family in Figure 3.9 which demonstrates the influence of size ratio, r, a key parameter determining the position of the diagram within an overall triangular envelope. Each voids ratio diagram consists of six key points, $A-F$ joined by five straight lines.

The extreme points are determined by the voids ratios U_0 and U_1 of the coarse and fine components respectively.

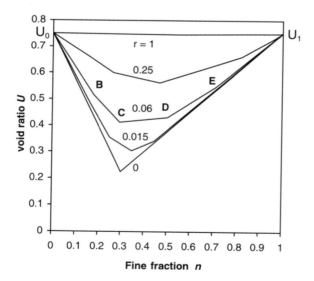

Figure 3.9 A set of theoretical voids ratio diagrams illustrating the effect of size ratio r on the resultant voids ratios of mixtures of two materials having the same voids ratio of 0.75.

The co-ordinates of the four intermediate points $B-E$ in a voids ratio diagram may be calculated for a mixture of two materials from the mean size and voids ratio of each material by the use of the following equations and Tables 3.1 and 3.2 repeated from earlier in the text

$$U_0'' = (1 + U_0)(1 + mr)^3 - 1 \tag{3.8}$$

$$Z = k_{int} + [(1 + U_0)^{1/3} - 1 - k_{int}]r^{k_p} \tag{3.15}$$

$$U_1'' = \frac{(1 + U_1)U_0''}{(1 + U_0'') - (1 + Z)^3} - 1 \tag{3.17}$$

Proportion of the finer material

$$n = \frac{U_0''}{(1 + U_1'' + U_0'')} \tag{3.18}$$

Table 3.1 Summary of values for spacing factors, m, for coarse particles in a mixture with fine particles

Point in voids ratio diagram	Spacing factor, m
A ($n = 0$)	0
B	0.3
C	0.75
D	3
E	7.5
F ($n = 1$)	∞

Table 3.2 Summary of values of empirical factors, for determining Z, the notional width factor of voids at each change point

Point in voids ratio diagram	k_{int}	k_p
B	0.12	0.60
C	0.06	0.65
D	0.015	0.8
E	0	0.9

Voids ratio of the mixture

$$U_n = nU''_1 \tag{3.19}$$

Voids ratios for intermediate points may be calculated by linear interpolation.

The voids and solids contents of the mixture at each main point or intermediate point may then be calculated from the following equations:

Voids content of the mixture

$$V_n = \frac{U_n}{1 + U_n} \tag{3.20}$$

Solids volume content of the mixture

$$S_n = \frac{1}{1 + U_n} \tag{3.21}$$

The solids volume content may be divided into coarse and fine contents in the proportions of $(1 - n)$ and n and may be converted to mass using the appropriate relative densities.

3.2 Illustrations of the main mechanisms and interactions within particle mixtures

The two main features of mixtures of two materials, having different mean sizes of particles, are

1. The interference of each size with the other resulting in reduction of the packing density of both
2. the smaller particles filling voids between the larger particles increasing the packing density overall.

Fortunately, if $r < 1$, there is usually an overall benefit, because the overall packing density always exceeds the proportionate density of the combination.

A subsidiary adverse feature is the possibility of segregation of the sizes which may occur because of the method of mixing, placing and compacting of the mixture and the relative sizes of the particles. In the modelling, this has been assumed to be negligible.

The features apply to graded materials as well as to mixtures of single sizes but for ease of presentation, only mixtures of two sizes are considered.

3.2.1 Particle interference in dry particle mixtures

Fine particles interfere with the packing of coarse particles. Coarse particles interfere with the packing of fine particles. Both are aspects of particle interference resulting in an increased voids ratio. The two aspects are illustrated by Figures 3.10 and 3.11.

In Figure 3.10, the coarse material structure is dilated by the introduction of fine material and the fine particle structure is similarly disrupted by the coarse particles. The voids ratios of both materials have thus been increased, although the resultant voids ratio of the mixture may be lower than either material considered separately.

Figure 3.11 illustrates how the introduction of a single coarse particle disrupts the fine particle structure requiring the displacement of a greater volume of fine material, and its associated voids, than the volume of the coarse particle. In addition, the fine particle structure may be affected adversely to some appreciable distance from the surface of the coarse particle.

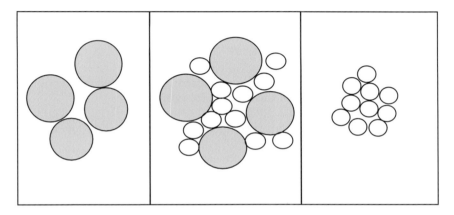

Figure 3.10 Illustration of the influence of particle interference on both the fine and coarse particle structure in a mixture of two sizes of materials.

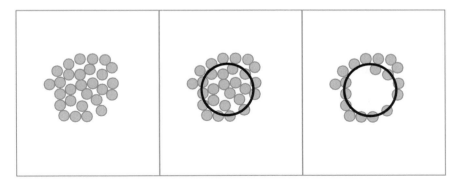

Figure 3.11 Illustration of the effect of particle interference due to introduction of coarse material into a 'fine particle dominant mixture'.

It will be observed that in the caption to Figure 3.11 reference is made to a 'fine particle dominant mixture', which is a description favoured by Powers (1968) together with its opposite, i.e. coarse particle dominant mixture. Generally, the author has tended to avoid the use of these terms because, while it might be reasonable to apply the latter description to $A-C$ in the voids ratio diagram and the former to $D-F$, it is less easy to decide a description for $C-D$ without needing to coin an intermediate description.

When the two components consist of graded materials it is assumed that the effects can be simulated in terms of two single sized materials having the mean sizes of the two components but having the voids ratios of the graded materials.

3.2.2 *The significance of the change points in voids ratio diagrams*

In considering the various features of voids ratio diagrams, the explanation is relatively complex because of the interaction between a variety of factors; these are:

- The voids ratios U_0 and U_1 of the coarse and fine components of the mixture respectively and their mean size ratio r
- The fine fraction, n
- Interference of the coarse component in the packing of the fine component
- Interference of the fine component in the packing of the coarse component
- The spacing factor m of the coarse component
- The ability of the fine component to fill or partially fill space between coarse particles when the coarse particles are relatively far apart and the limited extent to which filling can occur in narrow spaces between coarse particles when they are relatively close together

It is a practical frustration, and an apparent anomaly, that the point or zone of minimum voids ratio of a combination is to be found in the region where the effects of particle interference are greatest. The simplest explanation is that in using one material to fill the voids in another, it normally follows that the fine fraction n for minimum voids is remote from both 0 and 1.

Also, if both of the postulated types of particle interference have significant effects then the maximum composite effect is also likely to be at an n value remote from 0 and 1. Thus maximum detrimental effects of particle interference tend to occur in a similar zone to that associated with maximum benefit of void filling. However, there are exceptions due to particular properties of the materials, so that generalisation should be avoided.

The question arises as to why change points are associated with significant changes in slope in the voids ratio diagrams and with particular values of m, the coarse material spacing factor.

An attempt to illustrate the answers is provided in Table 3.3, Figures 3.12 to 3.15 and in the following discussion. The same example is considered as was covered by Figures 3.5 to 3.7 so that U_0, U_1 and r are constants.[9]

Considering first the AB zone when the fine fraction, n, is low, the fine material exists at a low density in only partially filled voids and particle interference is minimal but increasing towards B. At B, there is just sufficient fine material to fill the voids but at a low density due to the relatively small

Table 3.3 Description of the main features of particle interference and effects on the voids ratio diagram

Change point or zone	Spacing factor m	Main features	Interference
A	0	All coarse particles	
AB		Fine particles partially fill voids between coarse particles. Fine particles interfere only slightly with packing of coarse particles.	Minor interference
B	0.3	Just sufficient fine particles to fill voids between coarse particles without major dilation of the coarse particle structure.	
BC		Major interference in the packing of both fine and coarse particles. Insufficient gap between adjacent coarse particles to allow fine aggregate except at a very high voids ratio locally. When the fine particles do exist in these gaps, the coarse particle structure is further dilated locally.	Major interference
C	0.75	Coarse particle spacing at the lower limit for permitting fine particles to exist in the gaps between adjacent coarse particles.	
CD		Major interference in the packing of both fine and coarse particles. Sufficient gap between adjacent coarse particles to permit fine particles but at a locally high voids ratio.	Major interference
D	3	Coarse particle spacing at upper limit for major interference locally in the gap between adjacent coarse particles.	
DE		Minor interference with fine particle packing throughout.	Minor interference
E	7.5	Coarse particle spacing at the upper limit for influencing the overall fine particle structure.	
EF		Minor interference with fine particle structure only locally near the few coarse particles.	Minor local interference
F	$\propto$	All fine material.	

B

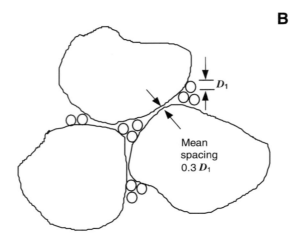

Figure 3.12 Coarse and fine particle distribution at change Point B for size ratio
$r = 0.085$

C

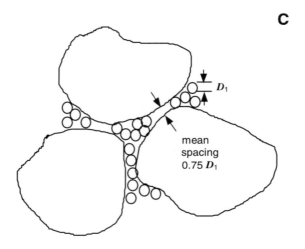

Figure 3.13 Coarse and fine particle distribution at change Point C for size ratio
$r = 0.085$.

voids and the presence of closely spaced boundaries formed by the coarse
particles.

In the *BC* zone, the coarse particles progressively move further apart to
permit more fine material with reduced interference, but, only at point C and
beyond is there scope for fine material to commence to exist at a reasonable
packing density in the closest spaces between the coarse particles. In the *CD*

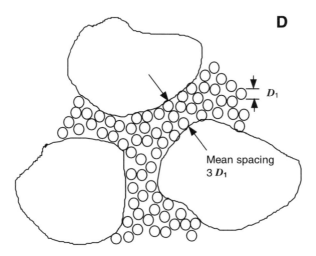

Figure 3.14 Coarse and fine particle distribution at change Point D for size ratio
$r = 0.085$.

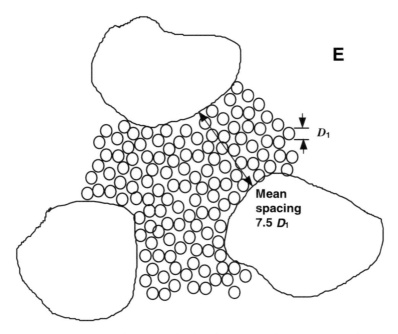

Figure 3.15 Coarse and fine particle distribution at change Point E for size ratio
$r = 0.085$.

zone, these spaces progressively widen until D is reached, when there is now sufficient space between coarse particles for generally improved packing and lower interference. In the DE zone there is progressively less interference until at E the coarse material interferes only near to its own interface with the fine material. Finally, as F is approached the amount of interference is small because there is now only a small fraction of coarse material to provide interference with the fine material in the immediate vicinity.

3.3 General theory of particle mixtures developed for mixtures of two graded components

The theory and formulae developed for mixtures of two single-sized components can be applied to graded components by calculating the mean size of each material as the proportionally weighted mean size on a logarithmic basis to account for the size distributions of each graded material, as described in Section 2.1.1.

Alternatively, if the voids ratios are known for each size fraction, the composite voids ratio of each component material and of resulting mixture can be calculated as for mixtures of multiple components, as in the following section.[10]

3.4 General theory of particle mixtures developed for mixtures of three or more components

The theory and formulae may also be applied to a mixture of more than two components by firstly using the formulae to calculate the mean size and voids ratio of the combination of the two finest materials, and then combining this combination with the next coarsest material and so on until all materials have been utilised and the resulting voids ratio of the total combination is obtained.

When the sizes of some components are reasonably close and it is intended to fix their proportions, it is technically justifiable to either

- Use the theory to calculate the composite mean size, voids ratio and relative density before combining the composite material with the other components
- Determine the composite value for voids ratio by laboratory tests

Either approach could be used for dealing with, for example, four materials by combining two sizes of coarse aggregate and two sizes of sand. The resulting two materials, i.e. a composite coarse aggregate and a composite fine aggregate, could then be blended together as described in the previous section for two materials.

Alternatively, any number of components can introduced by successively combining the finest with the next coarsest material and so on until all have been introduced.

For example, the case study in section 8.4 demonstrates the successive combination of up to 14 materials, and the computerised adjustment of the proportions of each of the components for minimum voids ratio.

3.5 Situations requiring special care or treatment

It is important to consider the sequencing of adding material in the simulation and to consider whether the gradings of individual materials may introduce problems.

The model is based on the sequence of combining the finest materials before adding the next coarsest material. When three or more materials are to be combined, the model may produce a higher resultant voids ratio if the sequence is reversed by combining the coarsest materials first before introducing the finest material. It is thus normally essential to combine the finest materials first before adding coarser material except that, when some materials are close in size, they can be pre-combined as described earlier.

The theory assumes that a graded material can be represented by a single-sized material having the same voids ratio as the graded material. However, problems may arise if a material with a gap in its grading is to be combined with a material with a mean size within that gap.

The problem may be overcome as indicated in the following example.

- Two materials A and B are to be combined.
- Material A has a substantial gap in its size distribution, such that it is effectively two materials A1, the finer and A2, the coarser. The voids ratios of A1 and A2 are known.
- The other material, B, having a size intermediate between A1 and A2 is to be mixed with A.
- The simulation will be inaccurate if A and B are combined directly. However, the difficulty can be overcome by first combining A1 and B and then adding the coarsest material A2 to the combination.

Another case, possibly requiring special consideration, is when a fine material is to be combined with coarser materials which have been pre-blended to produce a very wide grading. In which case it is recommended to separate the coarse combination into its components, to measure their separate properties and then to combine in the normal way, finest first and coarsest last. This situation would apply when combining a cement and an all-in aggregate consisting of pre-blended coarse and fine aggregates.

4 Extension of the theory of particle mixtures to pastes, mortars and concretes

4.1 Reference slump for mortars and concretes

The commonly selected value for concrete trials and for reference purposes by the ready mixed concrete industry is 50 mm slump. Cement pastes of standard consistence in the Vicat test produce slumps in the order of 50 mm. Mortars and concretes of 50 mm slump simulated by the theory do not need major adjustment to account for aggregate contents estimated from loose bulk densities. For these reasons, 50 mm was selected as the reference slump for simulation of pastes, mortars and concretes. Adjustments for other slumps are dealt with in section 4.4.3.

4.2 Pastes: extension of theory of particle mixtures to cement–water pastes and other powder–water pastes

A detailed examination of the correlation between cement and concrete properties, particularly water demands, is provided in Appendix C.

As a result, it is concluded that mixtures of powders in water can be simulated using the same techniques, formulae and values of constants as for dry particle mixtures on the basis of the mean sizes, voids ratios and relative densities of the powders, except that the voids ratios are determined in water using the Vicat test.

4.2.1 Extension of theory of particle mixtures to composite cement in water paste and to cement–addition combination in water paste

For composite cements containing additions such as fly-ash or blastfurnace slag, either interground or blended with the cement, the voids ratio can be estimated from a Vicat test as for Portland cement.

For combinations of a cement and additions blended dry or combinations blended wet in the concrete mixer, the voids ratio of the combination can be estimated by either

• Making a Vicat test on the combination

- Making Vicat tests on each component and then using the Theory of Particle Mixtures to estimate the voids ratio of the combination.

It is also necessary to estimate the mean size and relative density of the combination.

Depending on the properties of the components, the resulting composite cement or combination may result in a lower water demand paste. However, it should be noted that when a lower water demand paste is achieved it does not necessarily follow that a lower water demand concrete will be obtained. This will depend upon the interactions between the properties of the composite cement or combination with the fine and coarse aggregates in mortar and concrete, which will be a function of the relative sizes, as well as the voids ratios and proportioning of all components, as discussed in the following sections.

However, there is substantial experience with composite cements and combinations to show that significant benefits can be obtained for pastes, mortars and concretes in reducing water demand by the incorporation of additions.

Wimpenny (c1995) recognised that the principles of packing apply to cements, ggbs and to combinations of powders and concluded that a wider psd of a particular sample of ggbs would be advantageous in reducing the water demand of concrete. Domone and Soutsos (1995) reported benefits to be derived by combining two or three cementitious components.

With regard to fly-ash, Hobbs (1988) explored the effects of fly-ash on water demand and strength of concrete. With high quality fly-ashes, reductions of 7–12% in water demand of concrete were reported when 30 or 35% pfa was used in the cement and fly-ash combinations at medium cement contents, but at high Portland cement contents negligible reduction in water demand was obtained. Butler (1993) reported satisfactory trials with a superfine fly-ash having a mean size of 4.6 μm (0.0046 mm) and very high specific surface.

With regard to limestone dust as a fine component of cement, Jackson (1989) concluded from joint industry research into limestone filled cements that it was important for the intergrinding to continue until an appropriate level of fineness was achieved to ensure a minimum water demand. Brookbanks (1989) reported that water demands were not generally increased significantly for correctly processed limestone-filled cements. El-Didamony *et al.* (1993) concluded that 5% calcined cement dust could be used beneficially with conventional blends of cement with ground slag or silica fume. Nehdi *et al.* (1996) showed technological and cost benefits of combining a fine limestone filler and condensed silica fume with a Portland cement.

With regard to silica fume, FIP (1988) identified that for concretes at a cement content of 100 kg/m^3, water demand reduced as silica fume was added but increased for cement contents above 250 kg/m^3. The Concrete Society (1990) advised use of high-shear mixing action to overcome agglomeration of silica fume particles or to use the silica fume in a slurry form with water. The Concrete Society considered that the higher water demand was due to early chemical reaction causing formation of a gel and that it was essential to offset this by appropriate changes to the aggregate grading and by use of high-range water reducers. Roy and Silsbee (1994) recognised the benefits of fine particles such as silica-fume as components of cement blends and also the necessity for dispersing agents. ACI (1996) reported that silica fume may have a mean size of 0.10 μm (0.0001 mm) and a specific surface of $100 \times$ typical cement value but the water demand of cement paste may be 5 or $6 \times$ typical cement value. The smaller size of silica fume was judged to overcome partially the higher water demand when mixed with cement in concrete due to the filling of voids between cement particles. However, a high water demand would usually occur in the concrete unless a water reducing admixture was also used. Reference was made to agglomeration of particles but the view was expressed that this can be overcome by the normal mixing action in the concrete production process. Gutierrez and Canovas (1996), in examining high-performance concretes with silica fume, concluded that water demands of cement and silica fume influenced the water demand of concrete. Neville and Aitcin (1998) stress that silica fume because of its fine size improves packing, particularly at the interface with aggregate. This requires more than 5% and less than 15% by mass of cement to obtain maximum benefit and cost effectiveness.

Jackson (1989) reported that ground gypsum, and also ground limestone, had a fineness of about 800 m^2/kg compared with ground cement clinker at 320 m^2/kg. Thus, it is to be expected from the Theory of Particle Mixtures that the water demand of cement paste can be modified by changing the gypsum content as is reported to be the case by Sumner *et al.* (1989).

Experience in using the Theory of Particle Mixtures with some of these materials confirms the benefits to be obtained. A few examples concerning fly-ash and ground granulated blastfurnace slag are shown later in section 6.3.3 and a case study relating to an ultra-fine material is given in section 8.1.

Some additions, such as silica fume, super-fine fly-ash or metakaolin are so fine that it may be difficult to secure an accurate measure of the particle size distribution and mean size. They may also agglomerate in water and an accurate assessment of the water demand in the Vicat test may be difficult without modifying the mixing method (see also section 8.1).

When the addition to be blended is coarser than the cement and not expected to act significantly as a cementitious component it may be more relevant to treat it as part of the fine aggregate or as an intermediate component between that of fine aggregate and cement.(see also section 3.5 concerning sequencing).

4.3 Mortars: extension of theory of particle mixtures to cement–sand mortars

The principle in extending the theory of particulate mixtures from aggregates to freshly mixed mortar is that mortar can be considered as sand in a cement

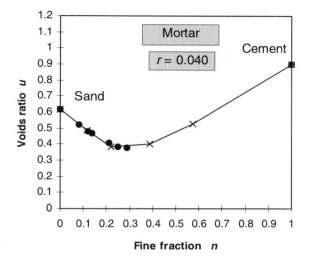

Figure 4.1 Voids ratio diagram for sand cement mortar mixtures showing good agreement between theory and experiment for a range of mortars of 50 mm slump (section 6.2).

paste matrix where the cement paste consists of cement particles in a water medium.

An example of a voids ratio diagram for mortars from section 6.2 is shown in Figure 4.1.

The values for N and U for each of the six points $A–F$ can then be converted to water, air and mortar solids contents by volume and finally to mass as described for aggregates in section 3.1.4.

4.4 Concretes: extension of theory of particle mixtures to concrete

The main principle is that concrete is a particulate material that may be simulated as coarse aggregate in a mortar matrix. Thus, a full range of mortars is first simulated as described in the previous section to determine the mortar parameters at the six change points. The simulation is then repeated using coarse aggregate with each of the six mortars (see later).[1]

This yields for each mortar a large range of concretes, only part of which is normally of practical use, e.g. some will have very high water demands, some will be over-cohesive and others will segregate. It is a function of the extension of the theory to identify the concretes having optimum proportions to resist segregation, avoid over-cohesion and to yield a low water demand. The concretes of practical relevance, and having these optimum properties, will normally be close to the minimum voids ratio on each voids ratio diagram.[2]

It had become clear from earlier work, Dewar (1986b) and Dewar and Anderson (1988 and 1992), through comparison of simulations and laboratory trials by experienced technical supervisors, that the conventional process of mix design and adjustment for safe cohesion of fresh concrete in the laboratory, could be consistently simulated, by ensuring that the selected value of n_X, the ratio of mortar solids to concrete solids at Point X, was about 0.025 higher than the value for minimum voids ratio (see for example Figure 4.3).[3]

This principle has been adopted generally throughout the simulations, but there are instances when other values than 0.025 are appropriate, e.g.

- A higher value might be necessary for concrete for pumping and for higher slumps than 100 mm or when the aggregates are more variable in their grading or shape.
- When experience with the concrete to meet local needs requires a higher level of cohesion.
- A lower value could be required when the gradings and void ratios of aggregates are modified during mixing and transporting through abrasion or degradation.

A cohesion adjustment factor CJ has been built-in to the calculation process for these purposes and the adjusted n value, n_X, is determined from equation 4.1.

$$n_x = n_{u\ min} + CJ \times 0.025 \qquad (4.1)$$

where

$n_{U\ min}$ is the proportion of mortar solids to concrete solids by volume at the point of minimum voids ratio

CJ is the cohesion adjustment factor, between 0 and say 3 or 4, normally 1.

The other co-ordinate, U_X, for each X point is determined assuming a straight line relationship between each pair of change points from a–f, as appropriate in the voids ratio diagrams. By implication, U_X is always slightly higher than the minimum voids ratio.

Thus, if point d has the lowest voids ratio then U_X is determined from

$$U_x = \frac{(n_x - n_d)}{(n_e - n_d)} \times (U_e - U_d) \qquad (4.2)$$

where

(n_d, U_d) are the co-ordinates of the lowest voids ratio

(n_e, U_e) are the co-ordinates of the point with the next highest n value
(n_X, U_X) are the co-ordinates of the intermediate point for safe cohesion.

Comparisons between predicted and actual void ratio diagrams for concrete demonstrate the necessity for adjustments to be made to voids ratios. This is compatible with the observation of Loedolff (1986) that 'wet' curves lay below the 'dry' curves for aggregate dominant mixtures but this reversed for powder dominant mixtures.

The reasons are not fully understood. They may be a reflection of the voids ratios at constant slump for wet concrete mixtures being affected differently from the model, which is based on dry mixtures of aggregates. There is a close correlation between the adjustment needed at any particular change point and the mean size of the mortar solids or mean size of the fine aggregate.[4]

The problem has been overcome in a practical way pro tem by the use of an empirical adjustment factor J, applied to the voids ratios of the concrete before conversion to water contents using equation 4.3.

$$J = k_1 - k_2 \times D_{FA} \tag{4.3}$$

where

D_{FA} is the mean size of the fine aggregate
   k_1 and k_2 are constants from Table 4.1 appropriate to the point $a-f$.

The adjusted voids ratios U_{JX} are calculated from the following formula for each point X

$$U_{JX} = \frac{U_X - J(1 + U_X)}{1 + J(1 + U_X)} \tag{4.4}$$

where U_X is the voids ratio at X before adjustment and J is the adjustment factor from equation 4.3.

Table 4.1 Voids ratio adjustment constants for the concretes at the six change points a–f for use in equation 4.3

Point	Voids ratio adjustment constants	
	k_1	k_2
a	0	0
b	0.0225	0.015
c	0.0225	0.0525
d	0.015	0.07
e	0.0125	0.07
f	−0.0425	0

The volumes (ssd) of materials at each point are determined from the

- Ratio of cement to mortar solids N
- Ratio of mortar solids to concrete solids n_X
- Adjusted voids ratio for concrete U_{JX}
- Per-cent air a

using the following formulae

Volume of cement

$$V_c = \frac{N.n_x}{1 + U_{jx}} \tag{4.5}$$

Volume of water

$$V_w = \frac{U_{jx}}{1 + U_{jx}} - V_a \tag{4.6}$$

Volume of fine aggregate

$$V_{fa} = \frac{n_x(1 - N)}{1 + U_{jx}} \tag{4.7}$$

Volume of coarse aggregate

$$V_{ca} = \frac{1 - n_x}{1 + U_{jx}} \tag{4.8}$$

Volume of air

$$V_a = \frac{a}{100} \tag{4.9}$$

It is finally necessary to convert from volume to the more conventional units of mass, utilising the relative densities of the materials.

4.4.1 Illustration of the process of applying the theory and models to concrete made with three materials

The process for extension of the theory to concrete consisting of three particulate components is demonstrated for a particular set of materials, Concrete Series 1- code F2.

1. Obtain test data for the materials and calculate (see section 2) the mean size, voids ratio and also the relative density for each material, as in Table 4.2.
2. For each change point *A–F* in the mortar voids ratio diagram as in Table 4.3 and Figure 4.2, determine *N*, the ratio of cement to mortar solids by volume, the voids ratio and the mean size of the mortar solids, e.g. at point *C* the value of *N* is 0.17, the voids ratio *U* is 0.32 and the mean size of the mortar solids *D* is 0.39 mm. The formulae are those for mortar which are also those for combining any two materials (section 3.1).
3. For each of the six mortars, *A–F*, calculate the concrete voids ratio diagram for the mixtures of mortar solids and coarse aggregate. (e.g. diagram for point *C* mortar shown in Figure 4.3).
4. For each of these six concrete voids ratio diagrams estimate the change point having the lowest voids ratio. (e.g. Point *d* in Figure 4.3).
5. For each diagram determine the intermediate point *X*, as 0.025 × cohesiveness factor to the right of the lowest point, providing a reasonable degree of safety against segregation, while still maintaining a low voids ratio (e.g. Point *X* in Figure 4.3, with a cohesiveness factor of 1.5 which results in an *n* value of 0.51, a voids ratio of 0.20 compared with values of 0.48 and 0.19 respectively for Point *d* in Figure 4.3). Figure 4.4 shows the diagrams for the six sets of combinations of coarse aggregate and mortars indicating the selected concretes at *X* in each diagram.[5]

Table 4.2 Mean sizes and voids ratios for Concrete Series 1 – F2

Material	Mean size (mm)	Voids ratio
Cement	0.015	0.83
Fine agg	0.75	0.51
Coarse agg	10.8	0.675

Table 4.3 Data calculated for the mortar voids ratio diagram

Mortar Point	Cement/cem + sand N	Voids ratio U	Mean size (mm) D
A	0	0.51	0.75
B	0.06	0.44	0.59
C	0.17	0.32	0.39
D	0.29	0.28	0.24
E	0.41	0.35	0.15
F	1	0.83	0.015

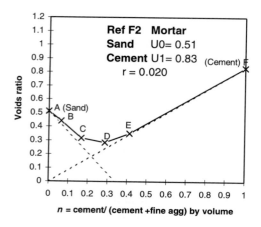

Figure 4.2 Example of a voids ratio diagram for mortar (Concrete Series 1 – F2).

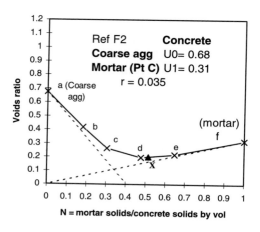

Figure 4.3 Example of one of the six concrete voids ratio diagrams for coarse aggregate and mortar C (from point C in Figure 4.2) (Concrete Series 1 – F2), illustrating the selected safety cohesive concrete X with low voids ratio.

6. Apply standardised adjustments (see equation 4.1) to the voids ratios of the 6 concretes to minimise residual errors, e.g. as shown in Table 4.4.[6]
7. Expand the simulated concretes by linear interpolation to provide, say, 15 or more, concretes to cover the range of potential production.
8. Convert the results for proportions and voids ratios for the six x points to conventional units as volume (m³) or mass per unit volume of concrete, as in Table 4.6, allowing for a residual air content, say 0.5 or 1%. In the example 1% has been adopted.

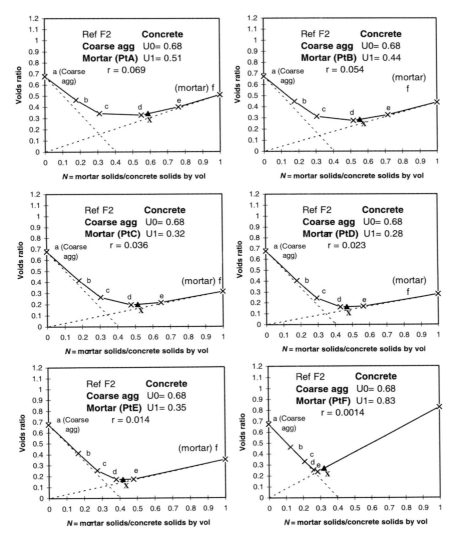

Figure 4.4 Voids ratio diagrams for the six sets of combinations of mortars and coarse aggregate showing the safe optimum combinations marked X (Concrete Series 1 – F2).

9. Prepare graphs to illustrate commonly required relationships as in Figures 4.5 to 4.7. Include laboratory trial data if available, for comparison and make minor adjustments if necessary. In the example, the assumed relative densities of the aggregates were increased by 20 kg/m^3.

Table 4.4 Example of output data for voids ratios and volumetric proportions from the voids ratio diagrams for the optimized concretes made with the six mortars A–F

Mortar		Concrete		
Ref	Cement/mortar solids by vol. Nx	Mortar/concrete solids by vol. nx	Voids ratio Unadjusted	Adjusted Ux
Ax	0	0.59	0.34	0.34
Bx	0.06	0.55	0.28	0.27
Cx	0.17	0.51	0.2	0.22
Dx	0.29	0.47	0.16	0.22
Ex	0.41	0.42	0.17	0.23
Fx	1	0.32	0.27	0.34

Table 4.5 Expansion of output data in Table 4.4 to cover 17 concretes (Concrete Series 1, F2)

Ref	nx	Ux	Nx
Ax	0.59	0.34	0
aabx	0.58	0.32	0.02
abx	0.57	0.30	0.03
abbx	0.56	0.29	0.05
Bx	0.55	0.27	0.06
bcx	0.54	0.26	0.09
bcx	0.53	0.24	0.11
bcx	0.52	0.23	0.14
Cx	0.51	0.22	0.17
cdx	0.49	0.22	0.23
Dx	0.47	0.22	0.29
dex	0.44	0.22	0.35
Ex	0.42	0.23	0.41
eefx	0.39	0.26	0.56
efx	0.37	0.29	0.71
effx	0.35	0.31	0.85
Fx	0.32	0.34	1

Finally, batch data can be prepared in conventional ways allowing for moisture contents and required concrete volumes and identifying concretes meeting particular specifications.

4.4.2 *Extension to multi-component concretes*

The extension to multi-component mixtures is merely one of adding as many

Table 4.6 Final output of 17 simulated concretes (Concrete Series 1, F2)

Cement	Water	Sand	Coarse	Density	Fines (%)	w/c
0	245	1116	768	2129	59.2	–
22	234	1098	795	2150	58.0	10.61
44	223	1080	823	2171	56.8	5.06
66	212	1062	852	2192	55.5	3.20
88	200	1045	882	2215	54.2	2.27
123	193	1006	909	2231	52.5	1.57
157	187	968	936	2248	50.8	1.19
191	180	931	964	2265	49.1	0.94
224	173	893	992	2282	47.4	0.77
292	171	794	1043	2298	43.2	0.59
352	168	699	1093	2313	39.0	0.48
405	173	600	1138	2316	34.5	0.43
448	177	509	1182	2317	30.1	0.40
560	195	352	1204	2312	22.6	0.35
650	212	216	1226	2303	15.0	0.33
719	228	99	1246	2292	7.3	0.32
768	244	0	1265	2277	0.0	0.32

Free water

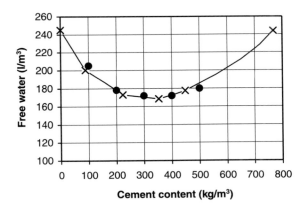

Figure 4.5 Example of a final water content diagram for concrete of 50 mm slump (Concrete Series 1 – F2).

intermediate stages as are required to the procedure described for three components.

The process can be simplified by fixing the proportions of some of the materials. These should be combined first, e.g. combining cement and fly-ash or combining two sizes of coarse aggregates, before moving to the main procedure for three components.

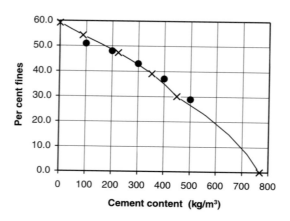

Figure 4.6 Example of a relationship between per cent fine aggregate and cement content (Concrete Series 1 – F2).

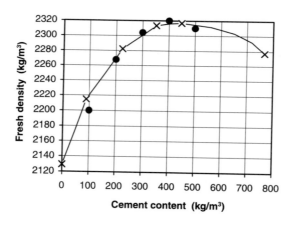

Figure 4.7 Example of a relationship between plastic density and cement content (Concrete Series 1 – F2).

4.4.3 Extension from the reference slump to other slumps

One means of adjustment is the following relationship adapted from one published by Dewar (1973) based upon laboratory data and published data in the literature.

Change in water demand

$$\delta W\% = 100(SL - RS)/6(SL + RS)$$

where SL is the intended slump and RS is the reference slump. e.g. If $SL = 75$ and $RS = 50$ then

$$\delta W\% = 3.3$$

If the original free water demand at the reference slump was 170 l/m^3 then the additional water required is

$$\delta W = 170 \times 3.3/100 = 5.7 \ l/m^3$$

The relationship is sensibly independent of materials and concrete parameters in the normal production range. Popovics (1965) has drawn similar conclusions in developing an alternative formula yielding similar results relating slump and water demand.

Using either method, or another accepted method, it is thus possible to adjust the water demands for any required slump.

The water demand adjustment factor F_S for slump can be calculated from

$$F_s = 1 + \frac{(SL - 50)}{6 \times (SL + 50)} \tag{4.10}$$

where SL is the intended slump in mm.

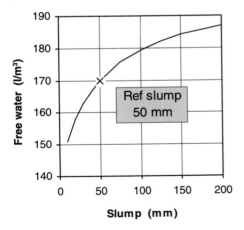

Figure 4.8 Example of the relationship between water demand and slump.

The water volume in equation 4.6 is adjusted correspondingly to

$$V_{wadj} = F_s \times V_w \tag{4.11}$$

Figure 4.8 shows an example using equations 4.10 and 4.11 and illustrates the influence of required slump on water demand, when the water demand is 170 l/m^3 at the reference slump of 50 mm.

In order to maintain the same volume of concrete at the changed water demand it is necessary to adjust the solids content. This can be done using the simple expedient of changing the coarse aggregate volume by the same amount as the water demand but in the opposite direction. This has the benefit that, when the water demand has been increased, the coarse aggregate content is reduced thereby increasing the cohesiveness to partly offset the loss of cohesion at the higher slump. The cement and fine aggregate contents are unchanged.

The coarse aggregate volume in equation 4.8 is adjusted to

$$V_{ca} = \frac{N \times n_x}{1 + U_{jx}} - V_{wadj} \times \frac{FS - 1}{FS} \tag{4.12}$$

5 Allowance for admixtures, air and other factors on water demand and strength of concrete

5.1 Influence of admixtures on the water demand of concrete

Admixtures may exert a major influence on the water and/or air void content of concrete by either physical or chemical means. Some may be employed deliberately for their water reducing or air entraining effects, while others may introduce such effects incidentally. The magnitudes of the effects may vary not only with the admixture dosage but also with such parameters as the cement properties, fine aggregate properties, the level of cement content and the workability of the concrete.

There are four obvious ways by which allowances could be made for the effects of admixtures, viz.:

- Start-point factors allowing for typical experience
- Tests of the materials to be used in the concrete
- Tests of concretes with the materials
- On-going modifications within a production/testing control system

In this work, only the first two options are under consideration with regard to effects of plasticisers on water demand and only the first option with regard to air entrainment. The third option formed the basis of assessment. The fourth option can be considered as continuous operation of the third option

As a result of the analysis of the data from Concrete Series 2, reported later in section 6.3.2 and the work of Austin (1995) it has been possible to suggest start-point admixture voids factors as shown in Table 5.1. The factor is applied as a **multiplying factor to the voids ratio of each material** to be included in the concrete. The factor is of course related to dosage rate.

Thus, when using a plasticiser, a start-point value of say 0.92 for the voids reduction factor might be applied, unless some higher or lower value is known from practice. It may be noted that, for a particular materials combination and admixture dosage, a value of 0.94 or 0.95 may be appropriate whereas a value of 0.88 was indicated from Concrete Series 2. This demonstrates the importance of having some prior data before deciding which values to use.

Table 5.1 Proposed start-point values for admixture voids factors for use with unknown materials and admixtures based upon Concrete Series 2 and the work of Austin (1995)

Admixture	Admixture voids factor
None	1
Plasticiser	0.88–0.95
Air entraining agent	1
Air entraining agent and plasticiser	0.88–0.95

For a high range water reducer or superplasticiser, a reduction factor as low as 0.80 or even 0.70 might apply.

ACI (1989) suggests that water reducing admixtures may reduce water demand by 5 to 8%, i.e. a reduction factor of 0.95 to 0.92, whilst for high range water reducers the reduction may be 12 to 25% or more, i.e. a reduction factor of 0.88 to 0.75 or less. These values are compatible with those proposed above.

Meyer and Perenchio (1982) in discussing the chemical influences of admixtures on the rate of stiffening of concrete identified the benefits from testing them in cement paste to assess the effects on water demand.

For the alternative option of testing of materials, a tentative proposal is made consistent with that of Meyer, and resulting from the work of Austin (1995), under the supervision of the author. The proposal is that the water reduction is measured for a cement paste in the Vicat test for standard consistence, first without the plasticiser and then with the plasticiser at the intended concrete dosage by mass of cement. The observed fractional reduction in the water demand, is then applied to the void ratios of *each* particulate material to be used in the concrete.[1]

In the particular work of Austin (1995), the water demand of the cement paste in the Vicat test was found to be reduced from 27.5 to 26%, i.e. a reduction factor of 0.945 by the use of the intended dosage of a particular plasticiser. The void ratios of the materials were reduced correspondingly with a similar resulting effect on water demand of the concrete.[2]

It will be seen from Table 5.2, that the predicted water demands of the concrete have been reduced by 10 l/m^3, i.e. also about 5.5%, by the inclusion of the plasticiser. The slumps are not substantially different from the intended value of 75 mm for either the plain concretes or those containing the plasticiser. All concretes were designed using the principles from this present work.

When other admixtures than plasticisers are incorporated in concrete it is necessary to consider whether they have any direct effect on water reduction and to account for it appropriately, as for plasticisers.

Table 5.2 Measured slump for concretes, with and without a plasticiser, designed by computer simulation for 75 mm slump (Austin 1995)

Admixture	Cement (kg/m³)	300	330	370	400
None	Water (l/m³) estimated for 75 mm slump	185	180	180	180
	Measured slump (mm) using the estimated water demands	80	80	85	85
0.6% by mass of cement	Water (l/m³) estimated for 75 mm slump	175	170	170	170
	Measured slump (mm) using the estimated water demands	80	75	80	80

5.1.1 Overview of allowance for the effects of admixtures on water demand

The plasticising effects of admixtures in reducing water demands of concrete may be simulated by either

- Testing the admixture at its intended dosage, with the cement to be used, in the Vicat test to determine the reduction factor for voids ratio and applying this reduction factor to the voids ratios of all materials to be used or
- Adopting start-point reduction factors from experience relating to the particular admixture and intended dosage (see also section 5.2.1 and Table 5.4)

The effects of admixtures on air entrainment are considered in the next section.

5.2 Air content

It has been assumed so far that the voids content of compacted fresh concrete is composed of free water and **entrapped** air so that, for a given total voids content determined by theory, a given workability and a given level of compaction as represented by a particular value for the percentage of **entrapped** air, the free water content may be determined very simply, by difference.

When **entrained** air is included by design or incidentally, the cohesive effect of the very small bubbles may be such that the corresponding reduction in water content may be less than the volume of entrained air. The actual effect may vary with the type of air entraining agent, its dosage, cement content and other factors which affect the quantity and size distribution of the entrained air bubbles. Some admixtures, e.g. plasticisers may also entrain air.

It is possibly logical that entrained air bubbles could be treated as particulate materials of zero density and included in the theory from considerations of their size distribution and shape, and to introduce corresponding values for mean size and voids ratio. However, although the size distribution is very occasionally determined from tests of hardened concrete, there are no standard tests for determining size distribution in fresh concrete. For this reason a theoretical approach has not been pursued, and instead an empirical approach has been adopted to account for entrained air. Examination of the literature provided the start point.

Figure 5.1, illustrating the data of Franklin and King (1971) and NRMCA (1977) shows that the water content can be expected to be appreciably reduced at lower cement contents by the introduction of entrained air, but that the benefit is considerably reduced at higher cement contents. It may also be seen that increasing the air content increased the water reduction, but not proportionally.[3]

If it is assumed that the air simply replaced water so that the total voids content was unchanged, then it would be expected that each 1% air would

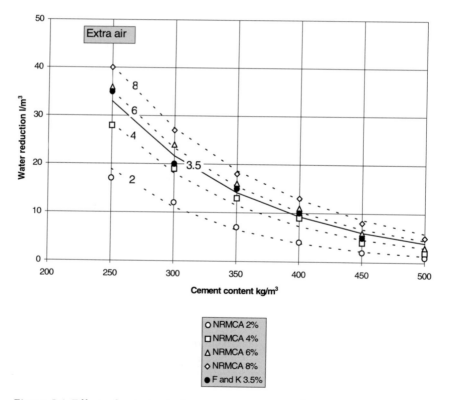

Figure 5.1 Effect of entrained air content on water reduction at different cement contents.

reduce water content by 10 l/m^3 whereas the actual reduction rates were always less than this value, even at the lowest cement content of 250 kg/m^3 and were much lower also as the air content was increased. Thus, the total voids content of air entrained concrete can be expected to exceed that of non-air entrained concretes except possibly at medium or low cement contents and low air contents.

In addition, the recent data for air-entrained concretes from Concrete Series 2 are shown in Figure 5.2.

An empirical relationship for water reduction WR was developed to provide the curves shown in Figure 5.1 and Figure 5.2, as follows

$$WR = k1 \times e^{\left\{\frac{c}{k2 \times a^{k3}}\right\}} \qquad l/m^3 \tag{5.1}$$

where $k1$, $k2$ and $k3$ are empirical constants, a is the entrained air content % and c is the cement content in kg/m^3.

The values of the constants for the data in Figure 5.1 and Figure 5.2 are shown in Table 5.3.

A comparison of the three sets of data in Figure 5.1 and Figure 5.2 is provided in Figure 5.3 based on the use in equation 5.1 of the constants in Table 5.3 when the entrained air content is 4.5%.

It will be seen from Figure 5.3 that the data from Concrete Series 2 and the corresponding constants from Table 5.3 are associated with lower water content reductions at the lower cement contents but higher reductions at high cement contents, compared with the earlier data from the literature. This may be due to the combined effects of the admixture formulations and the particular properties of the cements and fine aggregates used in the different

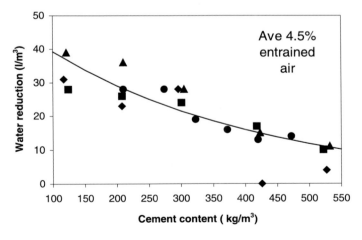

Figure 5.2 Relation between water reduction and cement content for air entrained concrete in Concrete Series 2.

Table 5.3 Values for constants in equation 5.1

Data Source	Constants		
	k1	k2	k3
Franklin and King	275	79	0.32
NRMCA	275	79	0.24
Concrete Series 2	53	135	0.60

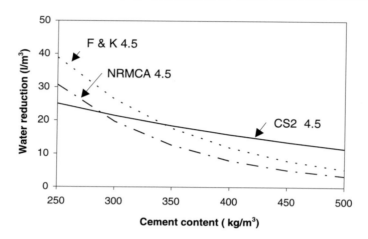

Figure 5.3 Comparison between the data from Figures 5.1 and 5.2 at 4.5% air estimated by the use of equation 5.1 and Table 5.3.

investigations. On average, reductions of 15–20 litres/m^3 are indicated for 4.5% entrained air.

Considering the overall picture, other workers have concluded similarly, e.g. Teychenne *et al.* (1988) assumed that a water reduction of 15 to 25 litres/m^3 applies for typical air contents obtained in air entrained concrete; ACI (1989) notes that normal air entrainment can reduce water demand by up to 10%, say 15 to 18 l/m^3 ; Gaynor (1968) reported that effects on mixing water requirement varied with aggregates as well as with cement content and admixture dosage.

5.2.1 Start-point values for effects of plasticisers and air entraining agents on water demand.

Start-point values for the parameters are suggested in Table 5.4 on the basis of section 5.1, section 5.2 and section 6.3.2.

In Table 5.4, the effect of entrained air on cohesion has been allowed for by reducing the cohesion factor G by $x/5$, based on an analysis of Table 6.29.

Table 5.4 Proposed start-point values for factors for use with unknown materials and admixtures

Admixture type	Factors							Air content (%)		
	Admixture voids factor	Air factors (for water demand)			Cohesion factors			Entrapped	Entrained (x)	Total
		a	b	c	Overall factor incl. air	Air factor (approx.)				
None	1				G	0		E	0	E
Plas	0.88	53	135	0.6	$G-x/5$	-0.2		E	F	$E+F$
AEA	1.00	53	135	0.6	$G-x/5$	-0.8		E	$T-E$	T
Plas/AEA	0.88	53	135	0.6	$G-x/5$	-1.0		E	$T-E+0.5$	$T+0.5$

Notes

E is normal entrapped air content.

T is specified intended total air content.

F is air entrained by plast'r.

x is the assumed entrained air content.

G is normal cohesion factor, (between 0.5 and 3, typically 1) for concrete without admixtures.

Approximate reductions in the factor are also shown in Table 5.4 for typical entrained air contents. If the adjustments are not made, an increased cohesion can be expected and may be adopted as a benefit.

5.3 Time and temperature effects

Examples of situations that may introduce time-dependent effects of particular significance for water demand for concrete were suggested by Dewar (1994) as follows:

- Fast stiffening cements
- Certain states of calcium sulfate in cements
- False or flash setting cements
- Clay minerals having high adsorption
- Some additives, admixtures or additions
- Relatively long times after mixing of concrete
- Relatively high ambient temperatures
- The use of dry absorptive aggregates
- Aggregates which degrade in the mixer
- Conditions of significant evaporation

Considering the effects of different cements, for example, in Figure 5.4, at the earlier time (or lower temperature) T1, cements C1 and C2 happen to behave similarly relative to one another in both paste and concrete with respect to water demand. At a later time T2, cement C2 out-performs C1. If the pastes

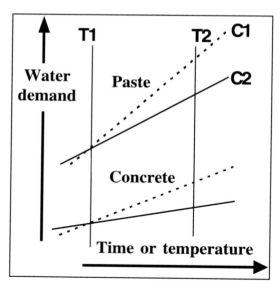

Figure 5.4 Time or temperature-dependent effects.

are tested at the earlier time and the concretes at the later time then the comparisons will indicate apparent anomalies.

To reduce the risk of problems of interpretation in such situations, as recommended by Austin (1995) for work in a hot country, the Vicat test should be made at a similar temperature and time after adding water to the cement as are required in practice for the concrete. Fortunately, experience may indicate that this recommendation can be relaxed for many normal situations and under temperate conditions.

Day (1995), also operating mainly in higher temperature climates and longer travel times for fresh concrete, introduced additional factors for time and ambient temperature in their effects on water demand, entrained air and cohesion (i.e. mix suitability factor of Day).

With regard to the other effects listed above, it is also essential to try to ensure that conditions in concrete trials mirror those in practice. When anomalies occur in making comparisons between theory, trials and practice it is recommended to consider whether any of the listed factors could be of relevance.

Sumner *et al.* (1989) for example refers to the influence of gypsum content of the cement on the water demand of the concrete, which is referred to again in Appendix C2.

As a further example, Brown (1993) stressed that some aggregates may disintegrate or degrade during mixing of concrete, so that aggregate tests may not be representative of aggregates in concrete, unless the tests are designed to detect such effects. Harmful clays, e.g. Illite, kaolinite or montmorillonite, in small quantities may have no detectable effect on the results of common aggregate tests but may have a marked effect on concrete water demand and strength.

5.4 Strength

5.4.1 *Relations between strength and w/c at 28 days*

In this sub-section, strength at 28 days only is considered, cement strength is assumed to be sensibly constant and effects of small amounts of entrapped air are discounted. Subsequent sub-sections deal with other ages and factors including air.

Water/cement ratio, or its inverse, cement/water ratio, has been recognised as a key influence on strength throughout the development of concrete technology.

A prime motivation for refinements has probably been the wish to transform curves or parts of them into straight lines for ease of drawing and comparing relationships. This has been achieved by using either a logarithmic scale for strength or by using c/w, or a power variant of c/w, rather than w/c. Hughes (1960) and Hobbs (1985), for example, preferred the Feret concept of relating strength to c/w rather than Abrams' (1924) preference for w/c.

Hansen (1992) developed a complex power law relating strength and w/c and accepted the Bolomey equation based on c/w as a working method. Popovics and Popovics (1994) assumed straight line relationships between strength on a log basis and w/c, in the range 0.37 to 1.0, using data from Kaplan (1960). Gutierrez and Canovas (1996) reported that, in Spain, c/w is favoured to enable use of a straight-line relationship with strength. However Gutierrez utilised the natural logarithm of strength for research and attributes different aspects of the relationship to cement and to aggregate. Baron *et al.* (1993) adopted a modified version of Feret's relationship between strength and c/w. Nagaraj *et al.* (1990) utilised log normal strength v w/c for analysis but later, Nagaraj and Zahida (1996) preferred strength v c/w.

Abrams (1924) established the well-known law of strength of concrete by relating strength, S, and the water/cement ratio, x, as

$$S = \frac{A}{B^x} \qquad\qquad (5.2)$$

where A and B are constants dependent on test conditions.

This may be adapted to a logarithmic basis as

$$LogS = A - x \times LogB \qquad\qquad (5.3)$$

A further need has been to take account of other factors, such as air content, age of test, cement strength, use of additions and admixtures, effects of aggregate type and to distinguish modifications judged to be needed at high strengths.

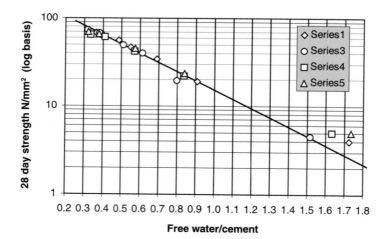

Figure 5.5 Example of a single straight-line relationship between strength (plotted on a log scale) and w/c (Concrete Series 2 – P1R1, P3R1, P4R1 and P5R1).

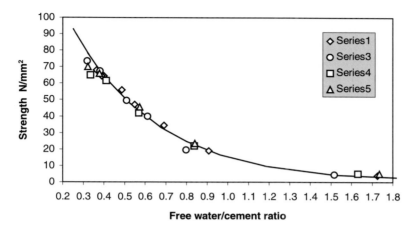

Figure 5.6 Example of a relationship between strength and w/c based on the single straight-line relationship of Figure 5.5.

A logarithmic scale for strength has been selected for the present work, on the basis of the author's experience over a number of years in dealing with these aspects, and in particular its ease of use when dealing with the following topic concerning changes in relationships at low w/c values.

Figures 5.5 and 5.6 demonstrate that, from Concrete Series 2 data for concrete strength at 28 days, a single relationship could be assumed conveniently for strength v w/c, over the range of w/c from 0.4 to at least 0.9. For higher w/c values, possibly a different relationship might apply.[4]

Sear *et al.* (1996) has investigated high w/c values in depth and for such situations reference should be made to his detailed work. Sear adopted Abrams' concept and the conversion to log basis for strength to obtain a straight line relationship with w/c. Sear also adjusted water demand to include air content. Sear found that a second straight line relationship was necessary at high water/cement ratios, above about 1.2, the change point being associated, according to Sear, with point B in the Dewar concept.

For w/c values below approximately 0.4, strengths progressively deviate from the general relationship. The effect, which is more apparent in Figure 5.6, is of practical significance and needs to be taken into account in the modelling process.

Hughes (1976), using c/w rather than w/c, has examined effects of low w/c values on relationships for concretes made with different aggregates and also identified that different aggregates produced relationships diverging systematically with reducing w/c. Hughes' data demonstrate that a change in relationship occurs in the vicinity of a w/c value of 0.4. Drinkgern (1994) illustrated a paper with an oft quoted diagram by Walz (1971) showing strength v w/c curves with a reversing of curvature at a w/c of about 0.50 for concretes made with four different cement qualities.

Assuming that a change in relationship does occur with all concretes, the question arises as to whether it is dependent on aggregate type or primarily a phenomenon associated with cement hydration and applicable to all test ages.

Nielsen (1993) suggested that, at a w/c value of 0.38, a change in relationship occurs for cement paste, because at lower values there is restricted space for hydration, as shown by Powers. However, Nielsen concluded by accepting 0.40 proposed by Hansen (1986), rather than 0.38 as the point of change.

If these workers are correct concerning a deviation at a w/c of about 0.40, then the effect is general and not directly associated with aggregate type and it should be manifest at all ages, being more apparent at lower values of w/c and at longer ages.

This is not meant to imply that aggregates do not influence strength but only that this particular phenomenon is possibly independent of the aggregate. The question of the effects of cement strength and type of aggregate are considered separately later in more detail although it will be observed that in Equations 5.4 and 5.5, factors are introduced for these aspects without immediate discussion.

To test the hypothesis concerning the change in relationship, the author examined Concrete Series 2 data and a wide range of other unpublished and published data. As a result of this work, it is postulated that a change in relationship does occur which can be conveniently simulated by assuming that for w/c values **below** 0.4, the intercept at zero w/c on the log-scale for strength is 0.6 × (intercept for w/c values **above** 0.4) for all the cement strengths, aggregate combinations and test ages examined. The compressive strengths in terms of cubes tested to BS 1881 may be calculated from the following formulae:

Free w/c $\geqslant 0.40$

$$f_{cu} = \frac{3.3\, f_{cem}}{10^{F \times w/c}} \times R\ \text{N/mm}^2 \tag{5.4}$$

Free w/c < 0.40

$$f_{cu} = 10^{\left\{ Log(P) - \frac{w/c}{0.40} \times Log(Q) \right\}}\ \text{N/mm}^2 \tag{5.5}$$

where

w/c is the ratio of free water to cement by mass

F is a composite strength factor for aggregate and age (section 5.4.5) $F = F_{age} \times F_{agg}$ $F_{age} = 1$ at 28 days

P $= 0.6 \times 3.3 \times f_{cem}$

$$\text{Log } Q = \text{Log } P - \text{Log } f_{0.40}$$

$R \qquad = 10^{(Ta)}$ is a composite factor for air (see later)

$f_{cem} \qquad$ is the cement strength at 28 days tested to EN 196

$f_{0.40} \qquad$ is the concrete cube strength at a free w/c of 0.40^5

In the case of the Concrete Series 2 data, the effect is to modify Figures 5.5 and 5.6 as shown in Figures 5.7 and 5.8.

5.4.2 Effects of cement strength on relations between concrete strength and w/c

Appendix C 1.2 provides support for the contention that concrete strength correlates with cement strength.

In Equations 5.4 and 5.5, F_{cem} is the cement strength in the EN 196 mortar prism test at 28 days.[6]

Somerville (1996) estimates that cement strength at 28 days in the BS 4550 concrete test has increased from about 33 N/mm^2 in the 1950s to about 45 N/mm^2 in the 1990s.[7]

Thus, in appraising older literature it is necessary to bear in mind trends in properties of cement and as well as changes to test methods.

Due to problems of converting between strength data from different countries both for cement strength and for concrete strength it is difficult to make direct comparisons. However, in cases where a particular research project has involved different cements, it has been possible to confirm the validity of the empirical approach of including an estimated cement strength in the calculation of the intercept in the log strength v w/c relationship.

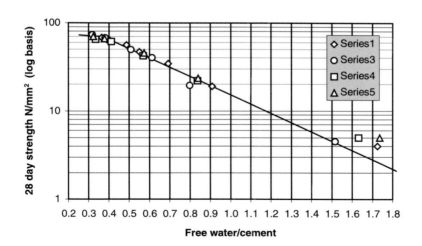

Figure 5.7 Example of a dual-straight-line relationship between strength (plotted on a log scale) against w/c for Concrete Series 2.

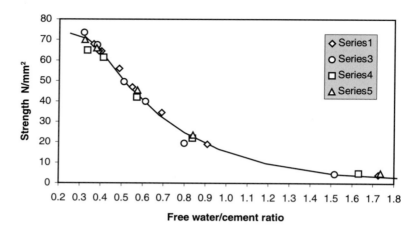

Figure 5.8 Example of a relationship between strength and w/c based on the dual-straight-line relationship of Figure 5.7 for Concrete Series 2.

For example, Mullick *et al.* (1983) and Visvesvaraya and Mullick (1987) published strength v water/cement ratio relationships for 6 cements having different strengths ranging from 35 to 60 N/mm² at 28 days to IS 4031–68. The data points, estimated from smooth curves of strength v w/c, are shown in Figure 5.9 together with relationships for each cement having intercepts of 2.97 × cement strength at w/c = 0 for w/c values of 0.40 and above and intercepts of 0.6 × 2.97 × cement strength for w/c values below 0.40. The difference between 2.97 in these calculations and 3.3 in equation 5.4 proposed for EN 196 tests may be due to differences in test methods and no significance is attached to it with regard to principle.

As discussed in Appendix C1.2, the work of Rendchen (1985) demonstrated a correlation between cement strength and concrete strength. Fagerlund (1994) also assumed a direct relationship between concrete strength and cement strength.

5.4.3 Effect of aggregates on relations between strength and w/c

Erntroy and Shacklock (1954) observed that strength v w/c curves for crushed granite and for irregular gravel diverged as the w/c reduced towards 0.3. Erntroy suggested the existence of 'ceiling' values for concrete strength for different types of aggregate.

Hughes (1976) demonstrated on the basis of strength v c/w relationships that concretes made with different aggregates (it is not clear whether the same cement was used) resulted in different relationships. Baron *et al.* (1993)

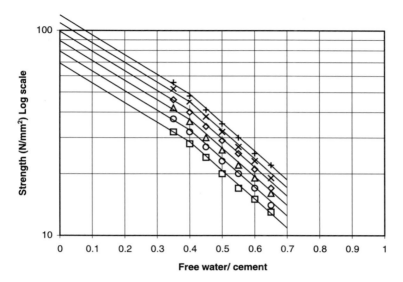

Figure 5.9 Influence of cement strength on concrete strength using data from Visvesvaraya (1987).

concluded that cement strengths and concrete strengths were related but that the relationship may not remain valid for very high strength concrete.

De Larrard and Belloc (1997) identified three parameters concerning aggregates which influence strength, namely bond strength, aggregate strength and 'maximum paste thickness' so-called, which seems to equate with mean spacing of coarse aggregate particles, higher strengths being associated with narrow spacing.

Uchikawa (c 1993) found that the ranking of aggregates in respect of concrete strength differed at a w/c of 0.2 compared with 0.5. Uchikawa suggested interfacial structure as the significant characteristic parameter. However, Hobbs (1976) considers that 'contrary to the widely held view, the compressive strength of concrete is not governed by the strength of the paste–aggregate interface'.

Examination of the data in these and other references suggested that for different aggregates, but the same cement and age at test, the relationships of log strength v free w/c had approximately the same intercept at w/c = 0 but different slopes. Taking the age factor F_{age} for slope as 1.0 at 28 days in equations 5.4 and 5.5, then the aggregate factor F_{agg} for slope could be determined directly from the slope of the data at 28 days (see also section 5.4.5).

For example, in Figure 5.9, for the tests of Visvesvaraya and Mullick (1987), F_{agg} was found to be a constant 1.4 for the 6 relationships at 28 days for different cements.

Observed values for F_{agg} varied for the data examined from 0.90 to 1.85, with the highest value applying to a combination of sintered fly-ash lightweight coarse aggregate and natural sand. A typical start-point value could be taken as 1.10.

A major research programme would be needed to determine whether any particular property or properties of the coarse and fine aggregates could be used to assess F_{agg} and whether other factors than aggregates are involved. For the time being. it is recommended that a start-point value for F_{agg} can be assumed as 1.10 or preferably determined from data for concrete made with the intended materials by testing strength at two or more values of free w/c spread apart in the range 0.40 to 0.90 (section 5.4.5). The value should then be monitored and adjusted continuously. If concrete is to be produced below a w/c of, say, 0.35, additional tests should be made to check that the value does not need fine-tuning.

Influence of maximum size of aggregate on strength v w/c relations

Erntroy and Shacklock (1952) found no effect of maximum aggregate size in the range 10 mm to 20 mm with two types of coarse aggregate. Walker and Bloem (1960) observed that decrease in maximum size in the range 60 mm to 10 mm of a particular aggregate resulted in a consistent increase in strength. Higginson in discussion of Walker and Bloem (1961) showed the same effect at low w/c but the opposite at high w/c. Bloem and Gaynor (1963) confirmed the earlier conclusions of Walker and Bloem (1960) over the range 40 mm to 20 mm. De Larrard (1997) emphasised the value of small maximum aggregate size in order to attain maximum strength.

Common experience in the UK indicates that strength increases as maximum size decreases such that F_{agg} could be reduced for smaller sizes. However, because different sizes of aggregates may be composed of different materials which may also influence F_{agg} no general recommendation is made and any modifications should be based on case evidence.

Influence of workability and aggregate content on strength v w/c relations

It is well recognised in the literature, e.g. Newman and Teychenne (1954), that at lower workabilities, higher strengths are obtained at a fixed w/c.[8]

It is also realised now, as discussed by Newman (1959, 1960) that most of the reported effect, occurring then in laboratory trials, was due to absorption of oven dried aggregate immediately prior to mixing with water in the concrete mixer. The effect is almost eliminated in practice by using wet aggregates but there may still be a real residual effect, which could be taken into account if desired, particularly when oven-dried aggregates are used for concrete trials without overnight pre-saturation. The author elected not to include a factor for this effect on the grounds that the reported concrete trials were made with saturated aggregates. In the case of cited references, it is not always clear what condition has been adopted for the aggregates and thus the author has not been able to account for the aggregate condition.

5.4.4 Relations between strength and w/c at other ages than 28 days

Most concrete specifications for strength relate to 28 days but some may specify other ages. For control purposes, most concrete producers rely on prediction of 28 day strength from early age tests. Thus, prediction of strength at different ages and conversion of strength at one age to strength at some other age are important aspects to be covered by modelling.

The most significant factor affecting the relationship between strength at different ages is the rate of strength gain of the cement. There are important models relating to cement chemistry but the input parameters are not directly accessible to the concrete industry and these have been discounted for the present purpose. Of greater direct relevance is the information on strength at particular ages available from cement test certificates and from standard tests to EN 196 of particular samples of cement used in trials.

In appraising the literature, it is necessary to take into account that cement properties may have changed. For example, Somerville (1996) quoting Corish indicates that the ratios of 3 : 28 day strengths and 7 : 28 day strengths have

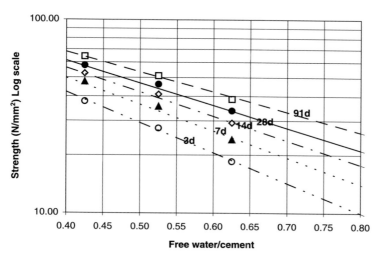

Ackroyd and Rhodes (1963)

Figure 5.10 Influence of age on concrete strength v w/c relationships for data of Ackroyd and Rhodes (1963).

changed from about 0.42 and 0.67 respectively to 0.55 and 0.75 between the 1950s and the 1990s.

Bearing this in mind, examination of cement test data and concrete data from the literature, and from recent practice, has enabled development of a relatively simple modification of the relationship between strength and w/c to allow for age of test by introduction of F_{age} in Equations 5.4 and 5.5 and assuming $F_{28d} = 1$. Figure 5.10 and Table 5.5, for example, show data obtained by Ackroyd and Rhodes (1963). The relationships intersect at about 178 N/mm^2 when w/c = 0.

Figure 5.11 shows examples of concrete strength v w/c curves for five ages from 1 to 91 days published by McIntosh (1966) for a rapid hardening Portland cement.[9]

In Figure 5.11 the intercept at A is 149, which is $3.3 \times$ EN 196 Cement Strength of 45 N/mm^2 at 28d and the intercept at B is 89, which is $0.6 \times$ A. The change point for slope is at w/c = 0.40 and the slopes of the relationships of log strength with w/c, for w/c values above 0.40, are as shown in Table 5.6.

The slope factors for age are related linearly to the reciprocal of the age on a log basis. Thus,

$$\text{Age factor} = \frac{\log 28}{((1-k) \times \log \text{Age} + k \times \log 28)} \tag{5.6}$$

and

$$k \cong 1/\text{Age factor at 1 day} \tag{5.7}$$

Table 5.5 Values of age factor F_{age} for data of Ackroyd and Rhodes (1963)

Slope factors					
Aggregate	Age				
	3d	7d	14d	28d	91d
1.16	1.35	1.19	1.08	1.00	0.90

Table 5.6 Slope factors in Figure 5.11

Slope factors					
Aggregate	Age				
	1d	3d	7d	28d	91d
1.1	2.07	1.43	1.23	1	0.86

McIntosh 1966

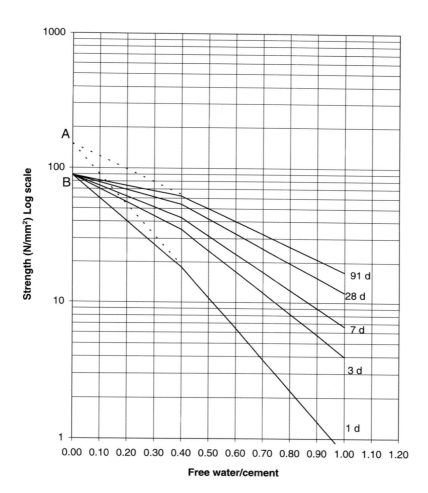

Figure 5.11 Example of the effects of age of test on the relationships between strength and w/c for concrete (McIntosh 1966).

in this instance k is 0.55 approx. The age factor is assumed to be unity at 28 days. The relationship is illustrated by Figure 5.12.

Using this relationship, and others relating strength and free w/c (Figure 5.11 and Equations 5.4 and 5.5), relations between strength and age may be determined as illustrated by Figure 5.13.

Another example of influence of age on relations between strength and free w/c is shown, for Concrete Series 2, in Figure 5.14.

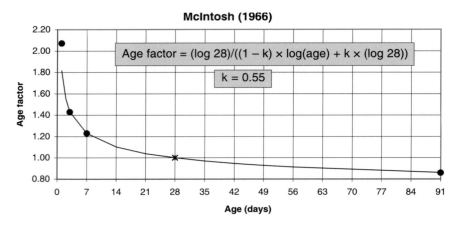

Figure 5.12 Relation between age factor for strength and age for data for rhpc concrete, from McIntosh (1966).

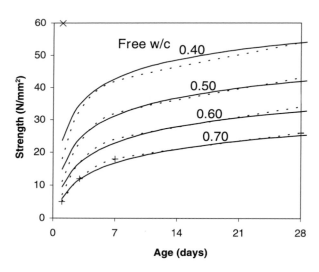

Figure 5.13 Comparison between empirical and observed (dotted) relationships between strength and age from data of McIntosh (1966) for concrete with rhpc.

In this case, the EN 196 cement strength is 51 N/mm^2 at 28 days and the slope factors are as shown in Table 5.7.

Relationships between 28 day strength and 3 and 7 day strengths for these data are shown in Figure 5.15; also shown is a typical point from cement test

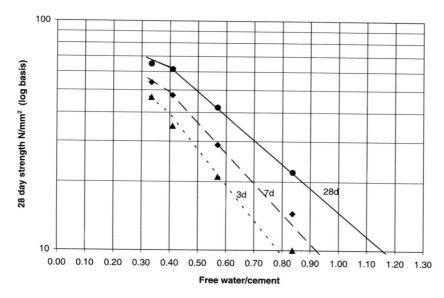

Figure 5.14 Example of relationships between 3, 7 and 28 day strength and w/c for Concrete Series 2 (P4R1).

data for 7 and 28 day strength. Figure 5.16 using the same data shows the gain in strength between 3 and 28 days and from 7 to 28 days.

It will be observed from Figure 5.16 that the modelling follows the test data in showing an increasing gain in strength from low to high strengths and that, towards the higher strengths, there is a levelling followed by a fall-off. This is associated with the change in slope of log strength at a w/c of 0.4.

5.4.5 Estimation of strength factors: f_{cem}; F_{agg}; F_{age}

Prior to obtaining relevant values from trial or production data, suitable start-point values for age factors as shown in Table 5.8 can be used in equations 5.4 and 5.5.

Table 5.7 Slope factors in Figure 5.14 for Concrete Series 2

Slope factors			
Aggregate	Age		
	3d	7d	28d
1.05	1.48	1.26	1

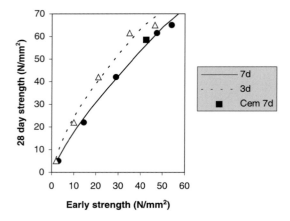

Figure 5.15 Relations between 3 and 7 day strengths v 28 day strength for Concrete Series 2 (P4R1).

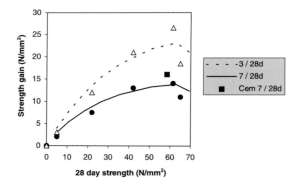

Figure 5.16 Example of relationships for strength gain from early to 28 day strength for Concrete Series 2 (P4R1). Included is a point for the typical cement strengths at 7 and 28 days for the source used.

Table 5.8 Start-point values for aggregate and age factors

Slope factors					
Aggregate F_{agg}	Age		F_{age}		
	1d	3d	7d	28d	91d
1.10	2.0	1.45	1.25	1	0.9

When trial or production data for concrete and data for cement strength are available the factors can be estimated more accurately. Two suitable methods for estimating concrete strength factors are

METHOD 1

Using 2 data sets for strength and free water/cement for non-air-entrained concrete.
Cement strength

$$f_{cem} = \frac{10^{\wedge}\{Log(f_{w/c2}) + [Log(f_{w/c1}) - Log(f_{w/c2})]/(1 - w/c1/w/c2)\}}{3.3 \times 10^{\wedge}(a \times (-0.038))} \quad (5.8)$$

Aggregate factor,

$$F_{agg} = [Log(f_{w/c1}) - Log(f_{w/c2})]/(w/c2 - w/c1) \quad (5.9)$$

Age factor,

$$F_{age} = [Log(f_{w/c1\ g}) - Log(f_{w/c2\ g})]/(w/c2 - w/c1)/f_{agg} \quad (5.10)$$

where

> $w/c1$ and $w/c2$ are the lower and higher free w/c values
> $f_{w/c1}$ and $f_{w/c2}$ are the corresponding 28 day strengths
> $f_{w/c1g}$ and $f_{w/c2g}$ are the strengths at an age of g days and
> a is the entrapped air content (%).

For example,

> $w/c1 = 0.465$ and $w/c2 = 0.87$
> $f_{w/c1} = 60.5$ N/mm^2 and $f_{w/c2} = 25$ N/mm^2 at 28 days
> $f_{w/c1\ g} = 51.5$ N/mm^2 and $f_{w/c2\ g} = 18.5$ N/mm^2 at 7 days
> $a = 1\%$,

resulting in the following estimates of the strength factors

> $f_{cem} = 55$ N/mm^2
> $F_{agg} = 0.95$
> $F_{age} = 1.15$ at 7 days

METHOD 2

Using cement strength (EN196 Mortar prism strength) at 28 days and 1 concrete data set for strength and free water/cement.

Aggregate factor,

$$F_{agg} = [Log(3.3 \times f_{cem}) - f_{cem}) - Log(f_{w/c}) - 0.038 \times a]/(w/c) \qquad (5.11)$$

Age factor,

$$F_{age} = [Log(3.3 \times f_{cem}) - Log(f_{w/c\ g}) - 0.038 \times a]/(w/c)/F_{agg} \qquad (5.12)$$

where

> f_{cem} is the cement strength (N/mm^2) at 28 days to EN 196
> w/c is the free w/c of the concrete
> $f_{w/c}$ is the corresponding 28 day concrete strength
> $f_{w/c\ g}$ is the corresponding strength at an age of g days and
> a is the entrapped air content (%).

For example, if

> $f_{cem} = 55$ N/mm^2 at 28 days to EN 196
> $w/c = 0.61$ for the concrete
> $f_{w/c} = 44$ N/mm^2 at 28 days
> $f_{w/c\ g} = 36$ N/mm^2 at 7 days
> $a = 1$ (%)

resulting in

> $F_{agg} = 0.95$
> $F_{age} = 1.15$ at 7 days

For Method 1, the two free water/cement ratios should preferably be spread apart, within the range 0.40–0.90. For Method 2, the single w/c should preferably be in the range 0.60–0.80.

5.4.6 *Relations between strength and cement content*

The relationship between strength and cement content at a stated consistence value is an important tool for concrete producers, because it relates a critical cost factor to a critical specification parameter, and takes account of the intricacies of product design for cohesion and specified consistence value, as well as the relationship between mean strength and w/c.

The expected relation takes the form of an 's' bend with a central section approximating to a straight line as described, for example, by Barber (1995, 1998). It is well known that different materials produce differing relationships that may converge, diverge or intersect so that prediction has been uncertain and recognised to be complex. The commonest solutions have been to adopt mathematical curve fitting techniques avoiding serious consideration of the technology involved. The ability to forecast effects is made more complex as the number of variables is increased, to include additions and admixtures, particularly plasticising and air entraining agents.

For these various economic and technical reasons, reliable simulation of the relationship between strength and cement content has become increasingly more of a necessity to avoid excessive time and effort in the laboratory. On the other hand, past effort in the preparation of trial concrete batches provides a wealth of data for validating simulation concepts.

By combining the relationship between water and cement content, derived from the Theory of Particle Mixtures, with the empirical relationship proposed between strength and w/c, it is possible to produce relationships that mirror concrete test data accurately as may be seen in Figure 5.17.

It will be observed that the relationship for strength v cement content in Figure 5.17 shows a distinct tendency to level off at high cement contents. Popovics (1990) identified a threshold for strength between 450 kg/m^3 and 500 kg/m^3 in cement content, but did not ascribe any reasons for it. Hughes

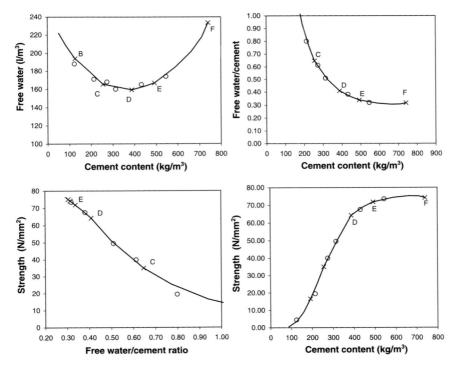

Figure 5.17 Illustration of the key relationships determining the more complex relationship between strength and cement content of concrete for Concrete Series 2 (P3R1) without admixtures; (open circles are trial data).

(1976) demonstrated on the basis of strength v c/w relationships that a change in slope occurred at a w/c in the range of about 0.35 to 0.45 as discussed in section 5.4.1.

It is suggested that the tendency is associated with two independent factors

- The change in gradient of the relationship between water content and cement content, associated with change points in the voids ratio diagram
- The change in gradient at about 0.40 in the strength v water/cement ratio relationship, associated with the chemical and physical aspects of hydration.

Additional examples, for concrete with admixtures, are provided in Figures 5.21 and 5.22.

5.4.7 *Influence of air on strength*

Following a long line of researchers, Hughes (1960) considered it important to include the effects of air voids when considering strength.

Early attempts to allow for the effects of air voids on strength did not distinguish air voids from water voids and more particularly did not take into account that the quantity of water voids reduced with age due to hydration whereas air voids were less affected. Popovics (1985) has discussed this in considerable detail and has separated the effects by the following modification to the historical formula relating strength to water/cement ratio

$$f = \frac{A_0}{B_0^{w/c}} \times 10^{-\gamma a} \qquad (5.13)$$

where f is strength, A_0 is a factor for the cement, B_0 is a composite factor for age at test and for aggregate, γ is 0.038 and a is the air content %. A version of this formula has been incorporated earlier into Equation 5.4.

Re-analysis of the data quoted by Popovics (1985), from a Hungarian paper by Ujhelyi (1980), yielded values of 180 for A_0, 20 for B_0 and confirmed the value of 0.038 for γ. Figure 5.18 compares predicted and measured strengths using this formula and the constants for the Ujhelyi data.

Figure 5.19 shows the effect of entrapped air in reducing strength, for the same data; the magnitude is similar to that indicated from the experience of Glanville *et al.* (1938), Kaplan (1960) and Stewart (quoted in Popovics (1985)) as shown in Figure 5.20.

The theoretical curves in Figures 5.19 and 5.20 have the following equation

$$Strength\ reduction\ \% = 100 \times (1 - 10^{-0.038 \times a}) \qquad (5.14)$$

In the view of Popovics (1985), based on assessment of data from Klieger, the formula is valid for entrained air as well as for entrapped air.

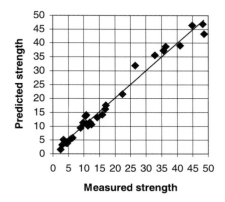

Figure 5.18 Comparison of measured and predicted strengths allowing for air content using equation 5.13.

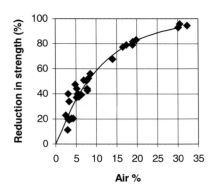

Figure 5.19 Effect of entrapped air in reducing strength.

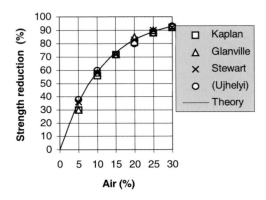

Figure 5.20 Effect of entrapped air content on strength reduction.

However, Wright (1953) reported that, at constant w/c, 5.5% reduction in air was obtained for each 1% increase in entrained air content up to 8%, compared with a reduction for entrapped air commencing at 7 to 8% reduction for each 1% air as shown in Figure 5.20. Similarly, Gaynor (1968) reports that air-entrainment in the range 0 to 10% added air resulted in a loss of strength of 5% per 1% air.

Teychenne *et al.* (1988) adopted a value of 5.5% reduction per 1% air for the design of air entrained concrete. The value of the constant in the formula would need to be modified from -0.038 to -0.025 to accommodate this lower rate of strength reduction.

Unpublished data made available to the author by Mr I. Smith of Fosroc Limited indicated from recent work in South Africa that, with admixtures in use today, the rate of strength reduction may vary with air content, necessitating the constant to be adjusted between -0.015 and -0.38 respectively for moderate and high (c 8%) values of entrained air content.

Thus, a more general formula is proposed as

$$\text{Strength reduction } \% = 100 \times \left(1 - 10^{(k_1 \times a_1 + k_2 \times a_2)}\right) \tag{5.15}$$

where

> k_1 is -0.038 for entrapped air
> a_1 is the entrapped air content (%)
> k_2 is -0.015 to -0.038 for entrained air; high values may apply to high percentages (c 8%) of air and in case of doubt and
> a_2 is the entrained air content (%)

Thus combining equations 5.13 and 5.15 leads to

$$f = \frac{A_0}{B_0^{w/c}} \times 10^{(k_1 \times a_1 + k_2 \times a_2)} \tag{5.16}$$

which allows for both entrapped and entrained air.

A reconciliation of the Concrete Series 2 data for air content, water demand, plastic density and strength indicated that an intermediate value for k_2 of -0.025 was appropriate for c 4–5% entrained air while maintaining the value for k_1 of -0.038 for entrapped air and high values of entrained air. A value for k_2 of -0.015 has been adopted for low contents of entrained air occasioned by the use of some plasticisers.

5.4.8 Composite effects of plasticiser and air entrainer on the relation between strength and cement content of concrete at constant slump

Figure 5.21 shows examples of simulated relationships compared with Concrete Series 2 data for strength and cement content of air entrained and

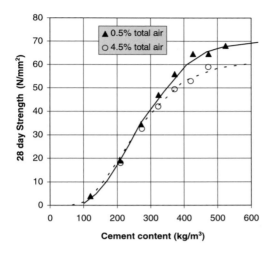

Figure 5.21 Example of simulation of the strengths of air-entrained and non-air entrained concretes in comparison with experimental data (Concrete Series 2 data P1 R3 and R1).

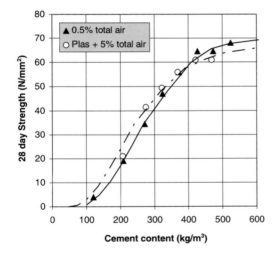

Figure 5.22 Example of a simulation of the strengths of plasticised air entrained and non-air entrained concretes in comparison with experimental data (Concrete Series 2 data P1 R4 and R1).

non-air-entrained concretes at constant slump. At low cement contents the reduced water content has offset the loss in strength due to air whereas at high cement contents, air increasingly dominates the relation. It is not surprising therefore that problems are sometimes reported of difficulty in meeting high strength specifications for air entrained concrete and why it is essential for accurate prediction of the effects of air. Figure 5.22 demonstrates simulation of the inclusion of a plasticiser to enhance further the performance of air entrained concretes.

5.4.9 *Effects of additions on strength*

Hobbs (1985), Teychenne *et al.* (1988), Hansen (1992) and Oluokun (1994) support the now traditional method proposed by Smith (1967), for additions such as fly ash, to be characterised with respect to strength by an efficiency factor, *e*, such that *a* kg of an addition could be assumed to be equivalent of $e \times a$ kg of Portland cement. Thus, the ratio $w/(c + e \times a)$ is used in place of w/c.

With some additions, the value of *e* may be quite low, e.g. zero at early ages rising to 0.25 at 28 days, with others it may be close to unity at 28 days and for a few the value may be very large e.g. 2 or more. For example, Hansen

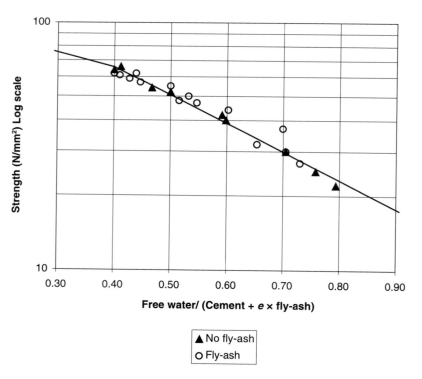

Figure 5.23 Relation between 28 day strength and water/(cement + e × fly-ash) for data from Hobbs (1985).

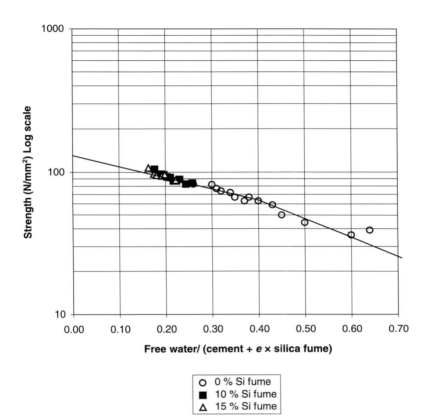

Gutierrez (1996)

Figure 5.24 Relation between 28 day strength and water/(cement + e × silica fume) for data from Gutierrez (1996).

(1992) recognised that for fly-ash, e varies with type of fly-ash, test age and curing. Teychenne *et al.* (1988) notes that for fly-ash, e may vary from 0.20 to 0.45 at 28 days and that for ground granulated blastfurnace slag, the factor can vary from about 0.4 to over 1.0. Fagerlund (1994) also utilises efficiency factor for additions, quoting typical values of 0.3 for fly-ash at 28 days and 0.8 to 1 for blast furnace slag. Persson (1998) demonstrated that the efficiency factor ranged from about 2 to 9 with higher values applying to low values of w/c, earlier ages and low percentages of silica fume.

Hobbs assumed a value of 0.25 for e at 28 days for a particular fly-ash and cement combination. The data obtained by Hobbs have been replotted as log strength v $w/(c + e \times a)$ in Figure 5.23.

Gutierrez and Canovas (1996) found that for a particular combination of cement and silica fume, the efficiency factor *e* for the silica fume was 4.75. Thus, if 10% of silica fume by mass of cement is added to a concrete having an original w/c value of 0.40, then assuming no change is required to the water content to maintain workability, the value of $w/(c + e \times a)$ will be $w/c/(1 + e \times a/c)$, i.e. $0.40/(1 + 4.75 \times 0.10) = 0.27$.

Thus, silica fume concretes can be utilised effectively to achieve very low equivalent water/cement ratios. For strength, silica fume can be accommodated within the Theory of Particle Mixtures in the same way as for fly-ash by the use the efficiency factor, as shown in Figure 5.24 for concretes with 10 or 15% silica fume when the efficiency factor has the value of 4.75 as determined by Gutierrez.

5.4.10 Overview of modelling for strength

Strength v w/c relationships can be modelled empirically by assuming a dual straight line relationship between log strength and w/c. The position and slope of the relationship is fixed by the cement strength and factors for age at test and aggregate. Modifications can be introduced to allow for the effects of entrained and entrapped air. By the use of an efficiency factor the effects of additions such as fly-ash, ground granulated blastfurnace slag and silica fume can be accommodated in the w/c. Strength v cement content relationships having the expected 's' bend shape can be simulated by combining the Theory of Particle Mixtures with the empirical relationships for strength.

6   Comparisons between theoretical and experimental data for aggregates, mortar and concrete

In this section, the results of laboratory tests of aggregates, mortar and concrete are compared with the results of computer simulations utilising the theory of particle mixtures, as follows

- Aggregate voids ratio diagrams   31 materials combinations of single sized or multi-sized aggregates
- Cement–sand mortar   6 trials of mortars
- Concrete Series 1   60 trials covering 12 materials combinations
- Concrete Series 2   92 trials covering 4 materials combinations with and without plasticising or air entraining admixtures
- Concrete Series 3   5 trials with fly-ash
6 trials with ground granulated slag

6.1 Aggregate mixture trials

6.1.1 Method

The following method applied to all experimental work with aggregates.

1. Bulk densities were measured in the oven-dry uncompacted (loosely poured) condition using a metal cylinder approx 0.007 m^3 capacity, generally following the method of BS 812: 1995.[1]
2. The bulk density of the coarser material was measured first and the quantity required to fill the container was then used for tests of mixtures.
3. An increment of the finer material was mixed with the coarse material and the bulk density of the mixture determined.
4. Further increments of the fine material were added and bulk density measured for each increment until the fine material content of the mixture exceeded 50%.

5. In Series JDD 1 the experimentation was repeated, but commencing with the finer material, and adding increments of the coarser material until the coarser material content exceeded 50%.
6. All masses were converted to volume using relative densities on an oven-dried basis and calculations made of the *n* and *U* values. Minor systematic adjustments were made to the *U* values over the range of data to eliminate experimental error at the 50% point associated with the separate halves of each set of experiments.

For convenience in reporting, single-sizes or half-sizes of aggregates have been coded as follows for the author's tests

Code	Nominal size range (mm)	Code	Nominal size range (mm)
A	14–10	b	1.18–0.6
B	10–5	c	0.6–0.3
x	5–2.36	d	0.3–0.15
a	2.36–1.18	e	0.15– 0.07

When relatively few points are obtained for density or voids, it is quite logical and justifiable to apply a smooth curve to illustrate the relationship, as appears to have been done, for example, by Furnas quoted by Powers (1968) and de Larrard and Buil (1987). However, the work of Powers (1968), Dewar (1983) and Loedolff (1985) showed the existence of straight line relationships for voids ratio diagrams for mixtures of aggregates, with clear points at which changes in slope occurred. The straight lines and change points are only seen clearly when sufficient closely spaced experimental data points are obtained.

For example, Table 6.1 shows the results for an experiment made, independently of the author, by Mr S.J. Martin of the RMC Group to determine whether the change points were real rather than imaginary and whether or not the lines were essentially straight. These results are plotted in Figure 6.1 and confirm the existence of change points *B–D* in the overall plot from *A–F* and also confirm the validity of assuming straight lines between the change points. Point *E* is usually, as in this case, the least distinct of the points and, when the size ratio *r* is large, *E* approaches *F*.

6.1.2 Observations

During experimentation, the following effects were observed for the tests involving small increments of the finer material, particularly when the size ratio was small. Neither effect is considered to have seriously affected the measurements.

Table 6.1 Data

Fines prop'n n	Voids ratio U
0	0.919
0.085	0.791
0.155	0.716
0.215	0.640
0.265	0.574
0.315	0.557
0.355	0.533
0.390	0.509
0.420	0.508
0.450	0.508
0.475	0.520
0.500	0.524
0.500	0.524
0.525	0.524
0.550	0.529
0.580	0.529
0.610	0.543
0.650	0.550
0.690	0.560
0.735	0.590
0.785	0.618
0.845	0.650
0.915	0.695
1	0.740

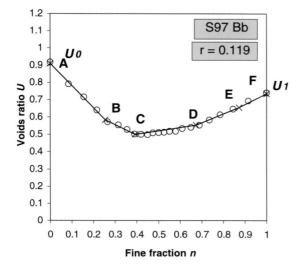

Figure 6.1 Voids ratio diagram demonstrating straight line relationships and change points.

Table 6.2 Series JDD1 – Experimental data for mixtures of single-sized aggregates

Materials code	Aa	Ab	Ac	Ad	Ae	Ba	Bb	Bc	Bd	Be	AB	ac	ad	bc
D mm	11.83	11.83	11.83	11.83	11.83	7.07	7.07	7.07	7.07	7.07	11.8	1.67	1.67	0.84
d mm	1.67	0.84	0.42	0.21	0.10	1.67	0.84	0.42	0.21	0.11	7.07	0.424	0.21	0.424
Adjustment factor F	1	1	0.85	1	1.5	1	1	0.6	0.6	1	1.3	1	0.8	1.3
Size ratio $r = Fd/D$	0.141	0.071	0.030	0.018	0.013	0.236	0.119	0.036	0.018	0.015	0.779	0.254	0.101	0.656

Prop'n of fine material n | Voids ratio U

n	Aa	Ab	Ac	Ad	Ae	Ba	Bb	Bc	Bd	Be	AB	ac	ad	bc
0	0.820	0.820	0.820	0.820	0.820	0.835	0.820	0.835	0.820	0.820	0.82	0.585	0.585	0.6
0.165	0.600	0.565	0.525	0.525	0.515	0.625	0.615	0.550	0.515	0.515	0.8	0.505	0.46	0.585
0.285	0.525	0.430	0.380	0.340	0.370	0.560	0.515	0.400	0.370	0.380	0.785	0.5	0.42	0.565
0.375	0.450	0.380	0.300	0.340	0.370	0.540	0.450	0.330	0.350	0.350	0.8	0.46	0.45	0.565
0.445	0.440	0.350	0.300	0.350	0.380	0.515	0.420	0.330	0.370	0.390	0.8	0.48	0.47	0.585
0.500	0.440	0.360	0.305	0.400	0.410	0.515	0.440	0.340	0.420	0.420	0.785	0.495	0.49	0.585
0.570	0.450	0.390	0.370	0.430	0.450	0.515	0.450	0.360	0.450	0.470	0.82	0.505	0.54	0.585
0.665	0.440	0.430	0.410	0.505	0.550	0.515	0.460	0.410	0.540	0.565	0.785	0.515	0.585	0.575
0.800	0.505	0.515	0.505	0.640	0.695	0.540	0.525	0.515	0.655	0.710	0.82	0.56	0.725	0.6
1	0.585	0.585	0.575	0.800	0.850	0.585	0.615	0.585	0.820	0.870	0.82	0.6	0.82	0.6

Table 6.3 Series JDD2 – Experimental data for mixtures of single-sized aggregates for *n* values of about 0.5 and below

Materials code	Aa	Ab	Ac	Ad	Ae	Bc
D mm	11.83	11.83	11.83	11.83	11.83	7.07
d mm	1.67	0.84	0.42	0.21	0.100	0.42
Adjustment factor F	1.2	1.2	1.2	1.2	1.2	1.3
Size ratio r = Fd/D	0.169	0.085	0.043	0.021	0.010	0.077

Prop'n of fine material n	Voids ratio U					
0.000	0.88	0.88	0.88	0.880	0.880	0.860
0.085	0.785	0.765	0.72			0.725
0.155	0.71	0.67	0.61			0.635
0.215	0.61	0.58	0.515	0.475		0.540
0.265	0.595	0.515	0.45	0.435	0.410	0.480
0.315	0.54	0.5	0.385	0.400	0.385	0.435
0.355	0.52	0.455	0.37	0.395	0.400	0.420
0.390	0.52	0.445	0.35	0.395		0.415
0.420	0.51	0.46	0.375	0.420		0.410
0.450	0.495	0.455	0.36			0.420
0.475	0.505	0.44	0.395			0.41
0.500	0.495	0.46				0.425
0.520	0.51					
1.000	0.635	0.655	0.6	0.880	0.915	0.6

1. Some segregation occurred after mixing of the fine and coarse material.
2. Some fine material filtered through gaps in the coarse material in the density container.[2]

6.1.3 Results of the main series of tests

Four series of tests were made as follows

- Series JDD1   Mixtures of single-sized aggregates
- Series JDD2   Mixtures of single-sized aggregates
- Series JDD3   Mixtures of multi-sized aggregates
- Series JDD4   Mixtures of typical fine and coarse multi-sized aggregates used in concrete.

The results are summarised in Tables 6.2–6.5 and Figures 6.2–6.5. In each of the figures, theoretical relationships are shown for comparison. These theoretical relationships are the final ones resulting from assessment of the constants discussed in section 3.1.[3,4]

Table 6.4 Series JDD3 – Experimental data for mixtures of two materials each having 1 to 4 components

Materials code	Ac	A abc	A xabc	Abx c	ABxa c	ABxa xabc
D mm	11.83	11.83	11.83	10.0	8.04	8.04
d mm	0.42	0.52	0.66	0.42	0.420	0.66
Adjustment factor F	1.2	1.2	1.2	1.2	1.2	1.2
Size ratio $r = Fd/D$	0.043	0.053	0.067	0.050	0.063	0.099

Prop'n of fine material n	Voids ratio U					
0	0.86	0.86	0.86	0.85	0.725	0.73
0.15	0.64	0.65	0.7	0.63	0.575	0.575
0.27	0.47	0.515	0.53	0.45	0.46	0.485
0.35	0.36	0.39	0.415	0.41	0.38	0.42
0.42	0.36	0.35	0.38	0.395	0.37	0.41
0.48	0.35	0.345	0.355	0.375	0.39	0.365
1	0.61	0.61	0.57	0.66	0.66	0.57

Table 6.5 Series JDD4 – Experimental data for mixtures of continuously graded fine and coarse aggregates

Materials code	G1	G2	G3	G4
D mm	11.48	11.48	11.48	11.48
d mm	0.97	0.62	0.450	0.36
Adjustment factor F	1.6	1.6	1.6	1.2
Size ratio $r = Fd/D$	0.135	0.086	0.063	0.038

Prop'n of fine material n	Voids ratio			
0.000	0.64	0.72	0.695	0.735
0.085	0.555	0.585		
0.155	0.495	0.515		
0.215	0.435	0.475	0.45	0.440
0.265	0.42	0.415	0.425	0.440
0.315	0.45	0.38	0.41	0.355
0.355	0.39	0.37	0.370	0.335
0.390	0.36	0.375	0.330	0.310
0.420	0.42	0.355	0.365	0.295
0.450	0.435	0.34	0.345	0.320
0.475	0.44	0.345	0.340	0.305
1.000	0.46	0.48	0.545	0.59

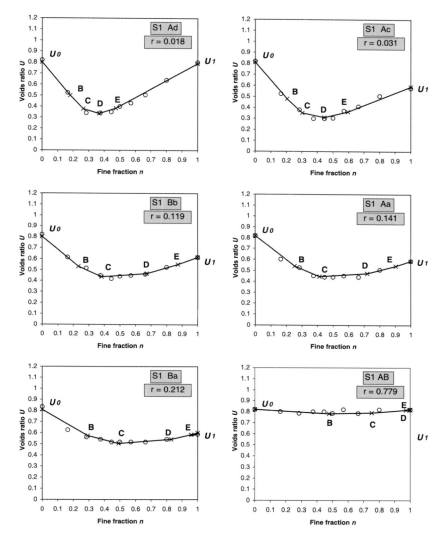

Figure 6.2 Examples from Series JDD1 of voids ratio diagrams for mixtures of single sizes of aggregates. Experimental data from Table 6.2 are plotted in comparison with theoretical relationships. The change points B–E are indicated by crosses.

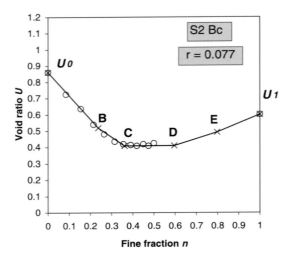

Figure 6.3 Example from Series JDD2 of the straight line relationships and changes in gradient in voids ratio diagrams.

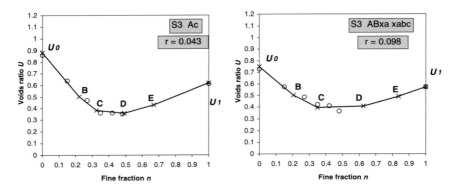

Figure 6.4 Examples from Series JDD3 comparing theoretical void ratio diagrams and experimental data for mixtures of two materials consisting of one or four components.

6.1.4 Overview of results from aggregate mixture trials

Comparison of the theoretical voids ratio diagrams with the experimental results for the four JDD series of tests of mixtures of aggregates, suggest that the behaviour of mixtures of single-sized and multi-sized materials, of a wide range of size ratios and voids ratios, in an uncompacted condition, can be predicted by the same theoretical formulae and set of constants.

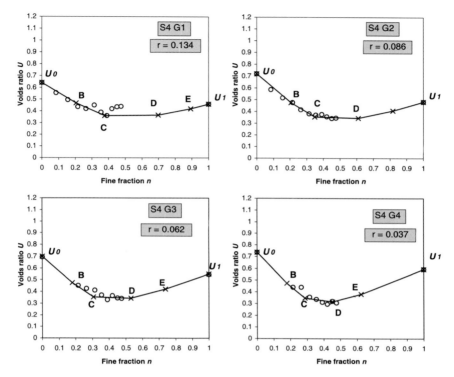

Figure 6.5 Series JDD4 – voids ratio diagrams for mixtures of continuously graded fine and coarse concreting aggregates.

The relative insensitivity of the normal range of sieve sizes resulted in inaccuracy in assessing the size ratio, which was overcome by adjustment of calculated size ratios on the basis of analysis of the voids ratio diagrams for minimum error.

6.2 Mortar trials

To test whether the theory of mixtures was also applicable to cement-sand mortar, a series of six trials was made with cement contents ranging from 175 to 665 kg/m^3 and water contents adjusted for 50 mm slump.

Voids content was calculated as the air plus free water content, and the fine fraction, n, was calculated as the ratio of cement to cement plus sand by volume. The air content was assumed to be 2% of the total mortar volume. The input data for the computer simulation, determined from tests of the materials, were as shown in Table 6.6.

Table 6.6 Materials input data for mortar

Materials data			
Material	*Mean size (mm)*	*Voids ratio*	*Relative density*
Cement	0.017	0.90	3.2
Sand	0.42	0.62	2.6

The measured results for *n* and *u* and the theoretical voids ratio diagram are shown in Figure 6.6. The close agreement between the experimental values and those generated by theory confirmed that the theory can be transferred from aggregates to cement-fine aggregate mixtures in water at 50 mm slump, without the necessity for adjustment.

The experimental results for cement and water contents are plotted in Figure 6.7 in comparison with the theoretical values based on 2% air. The closeness of agreement confirms that the water contents of mortars at 50 mm slump can be predicted directly from the theory.

It is assumed that water contents at other slumps than 50 mm can be factored in the same way as proposed for concrete in section 4.4.3.

6.2.1 Overview of results from mortar trials

The results of the cement–sand mortar trials suggest that mortars at 50 mm slump can be simulated as cement–sand mixtures using the same models, formulae and constants as for aggregate mixtures. The results for concrete in

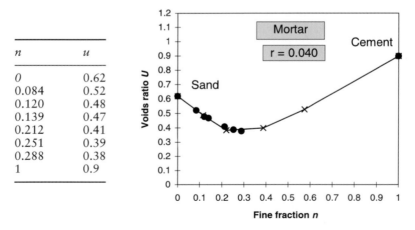

n	*u*
0	0.62
0.084	0.52
0.120	0.48
0.139	0.47
0.212	0.41
0.251	0.39
0.288	0.38
1	0.9

Figure 6.6 Voids ratio diagram for cement-sand mortar at 50 mm slump.

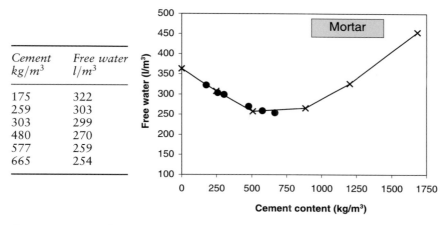

Cement kg/m^3	Free water l/m^3
175	322
259	303
303	299
480	270
577	259
665	254

Figure 6.7 Water demand of cement–sand mortar at 50 mm slump.

the next section provide additional validation, because mortar modelling is a prerequisite for proceeding to concrete modelling. However, to test the full validity of the extension to all mortars used for building purposes would require a more comprehensive study of a wider range of materials than considered necessary for the present project. In particular, it would be necessary to consider building sands rather than a concreting sand and to consider how to model the inclusion of slaked lime.

6.3 Concrete trials

6.3.1 Concrete Series 1 – covering a range of cement and aggregate sources

Comprehensive concrete trials were made with 12 combinations of materials from different locations in the UK. The experimental method was that described by Dewar (1986), in common use by the UK ready mixed concrete industry. The main features are

- Sufficient quantities are obtained of materials representative of current production
- Samples of materials are tested or certificates obtained from manufacturers.
- Aggregates are oven-dried and quantities for concrete trials are weighed and pre-saturated for 24 h with a known amount of water to allow re-absorption to simulate normal saturated condition in practice.

- Preliminary trials are made to determine the safe proportions of fine to total aggregates over the intended range of cement content to minimise water content, while maintaining adequate cohesion at 100 mm slump, the judgements being made by experienced supervisors or senior technicians familiar with the local concretes and materials.
- Concrete trials are made at five or more cement contents, in the range of cement content from 100 to 500 kg/m^3, at 50 mm slump, using smoothed values of per cent fines from the preliminary trials.
- Measurements are made of water content and density of the fresh concrete.

Table 6.7 Descriptions of the aggregates used in Concrete Series 1

Code	Coarse aggregate	Fine aggregate
B1	20–5 mm graded predominantly flint pit gravel. Irregular.	Siliceous sand. Irregular.
B2	20–5 mm graded crushed limestone. Angular.	50/50% crushed limestone fines and siliceous marine sand
F1	20–5 mm graded partially crushed flint pit gravel. Irregular.	Siliceous pit sand.
F2	20 and 10 mm single sized flint gravel combined 70/30% in the laboratory. Irregular.	Siliceous sand. Irregular.
G	20–5 mm graded crushed limestone. Angular.	Siliceous pit sand.
H	20 and 10 mm single sized oolitic limestone pit gravels combined 65/35% in the laboratory. Rounded/flaky.	Oolitic limestone pit sand. Rounded.
I	20 and 10 mm single sized crushed flint pit gravel combined 70/30% in the laboratory. Angular/rounded; elongated/flaky.	Marine dredged siliceous sand with high shell content. Rounded/ flaky.
K	20–5 mm graded predominantly flint pit gravel. Irregular.	Siliceous sand. Irregular.
N	20 and 10 mm single sized flint gravel combined 70/30% in the laboratory. Irregular.	Siliceous sand. Irregular.
P	20–5 mm graded, predominantly quartzite and oolitic limestone pit gravel. Irregular/flaky.	Siliceous pit sand.
T1	20–5 mm graded quartzite pit gravel. Rounded.	Siliceous pit sand.
T2	20 and 10 mm marine dredged flint gravel combined 70/30% in the laboratory. Rounded. High flaky shell content in 10–5 mm size.	Siliceous pit sand. Rounded.

Table 6.8 Concrete Series 1 – Reported or measured properties of materials including adjusted values

Materials code	Mean size (mm)			Voids ratio			Density (kg/m³)		
	Cement	Fine agg.	Coarse ag.	Cement	Fine agg.	Coarse ag.	Cement	Fine agg. ssd	Coarse ag. ssd
B1	0.015	0.77	10.48	0.9	0.64	0.86	3200	2720	2710
adj								2660	2650
B2	0.013	0.595	10.9	0.83	0.55	0.88	3200	2645	2690
F1	0.013	0.68	10.8	0.89	0.58	0.74	3200	2580	2520
adj								2590	2540
F2	0.015	0.75	10.8	0.83	0.51	0.675	3200	2550	2460
adj									2500
G	0.012	0.65	10.9	0.925	0.645	0.905	3200	2650	2690
adj				0.94	0.56	0.83			
H	0.013	0.88	10.6	0.925	0.53	0.71	3200	2440	2530
adj		0.9						2520	2610
I	0.014	0.54	11.2	0.925	0.71	0.85	3200	2660	2520
adj				0.9					
K	0.015	0.665	10.7	0.9	0.515	0.785	3200	2600	2570
adj	0.012			0.81					
N	0.015	0.73	11.1	0.875	0.55	0.605	3200	2580	2440
adj				0.85	0.5				2490
P	0.0135	0.31	11	0.89	0.59	0.765	3200	2540	2640
adj	0.012			0.87					
T1	0.013	0.47	10.5	0.925	0.65	0.635	3200	2590	2600
T2	0.013	0.45	11.6	0.96	0.7	0.69	3200	2630	2570
adj				0.99					

- Data are smoothed graphically and records made at cement contents of 100, 200, 300, 400 and 500 kg/m^3 for free water content, per cent fines and fresh density.

The combinations of materials have been coded to maintain commercial confidentiality.

The cements complied with BS 12 for ordinary Portland cement and were from different sources in the UK.

The aggregates complied with BS 882 and were from different sources in and around the UK. Summary descriptions are provided in Table 6.7.

Results for selected properties of the cements and aggregates are summarised in Table 6.8.

The adjusted values in Table 6.8 relate to adjustments made to minimise residual errors between predicted and observed values of water demand, per cent fines or fresh concrete density in Table 6.9, Table 6.10 and Table 6.11 respectively.[5-7]

The ability of theory to predict water demand of fresh concrete is demonstrated in Figure 6.8.

Detailed examination of Table 6.9 and Figure 6.8 shows that of 60 pairs of water demand values, 95% lie within ± 10 l/m^3 and 63% within ± 5 l/m^3 of the line of equality. It can also be seen that precision is poorer for the cement

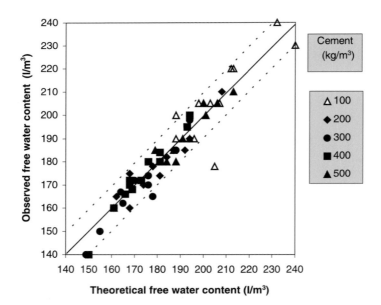

Figure 6.8 Concrete Series 1 – comparison between theoretical and observed water demands of fresh concrete of 50 mm slump made with different Portland cements and a range of aggregates of 20 mm maximum size. The dotted lines are drawn at 10 l/m^3 on either side of the equality line.

Table 6.9 Concrete Series 1 – Predicted and observed water demands of fresh concrete

Materials code	Water content (l/m³) Cement content (kg/m³)									
	100		200		300		400		500	
	Theory	Observed	Theory	Observed	Theory	Observed	Theory	Observed	Theory	Observed
B1	240	230	208	210	194	198	193	195	206	205
B2	205	178	181	174	169	172	168	172	179	185
F1	213	220	187	185	178	165	181	180	200	205
F2	198	205	178	178	170	172	173	172	188	180
G	207	205	184	182	176	174	181	184	201	200
H	212	220	194	190	188	185	194	200	213	210
I	232	240	192	185	176	170	176	180	187	185
K	196	190	174	170	165	162	166	166	181	180
N	188	190	168	175	164	167	169	168	188	180
P	188	200	161	160	149	140	150	140	161	160
T1	194	190	162	165	155	150	161	160	184	180
T2	203	205	168	160	161	160	168	170	191	190
Average	206	206	180	178	170	168	173	174	190	188

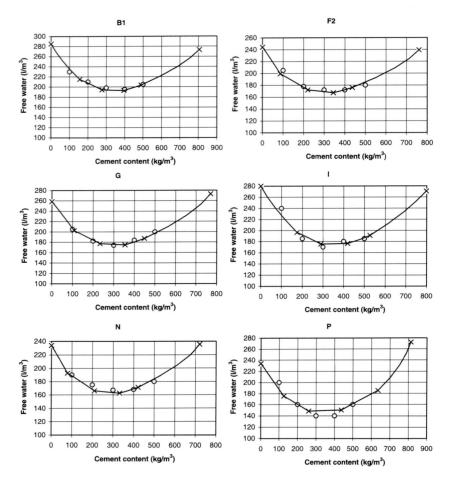

Figure 6.9 Concrete Series 1 – Examples of relationships between free water content and cement content for concrete of 50 mm slump showing comparison between theory and observation.

content of 100 kg/m^3 compared with higher cement contents. For the integrity of the theory, no significance is attached to this effect. It is common experience that, at low cement contents, slump and water content are more variable due to reduced cohesion and the consequent ease with which water can bleed very quickly during the slump test, often resulting in erratic test data.

Examples of theoretical relationships between water demand and cement content are shown in Figure 6.9 in comparison with observed values.

Table 6.10 Concrete Series 1 – Predicted and observed per cent fines of fresh concrete

Materials code	Per cent fines (Fine/total aggregate %)									
	Cement content (kg/m³)									
	100		200		300		400		500	
	Theory	Observed	Theory	Observed	Theory	Observed	Theory	Observed	Theory	Observed
B1	47	53	50	49	46	45	39	38	30	29
B2	52	51	48	48	44	43	39	38	31	29
F1	48	49	46	47	41	44	34	36	24	26
F2	54	51	49	48	42	43	35	37	25	29
G	52	48	48	46	42	42	36	37	27	31
H	55	54	49	48	42	42	33	36	23	29
I	41	52	45	47	42	42	37	35	29	26
K	53	53	48	49	43	45	36	40	28	35
N	51	48	46	44	39	40	30	36	21	28
P	43	44	41	41	37	38	33	34	29	28
T1	42	41	38	38	32	35	25	30	17	24
T2	42	41	38	38	33	33	27	28	18	21
Average	48	49	46	45	40	41	34	35	25	28

Cohesion factor is assumed to be 1, except for P, for which 2 was adopted, and for F2 and N, for which 1.5 was adopted (section 4.4.1 for explanation of cohesion factor).

There is very close agreement both on average and for individual materials combinations between theory and observation with regard to per cent fines. Of the 60 observations, 93% were within ±6% of theory and 80% within ±3%.

It will be apparent from Figure 6.10 that there is a tendency for the per cent fines values selected by the experienced supervisors to be higher than predicted for some of the leaner and richer concretes. No significance is attached to this with respect to the validity of the theory, for the following reasons

- The supervisors followed established practice by increasing the per cent fines values for the leaner concretes although it is clear from theory that with some materials combinations there is a maximum value for per cent fines beyond which no benefit is obtained; indeed the use of a higher value could in some cases affect water demand and cohesion adversely.
- Some supervisors were reluctant to use very low values for per cent fines in rich concretes, probably because of established local practice, so that in some cases there was a progressive additional over-sanding at higher cement contents.

This is considered further in examination of six examples of relationships between per cent fines and cement content shown in Figure 6.11.

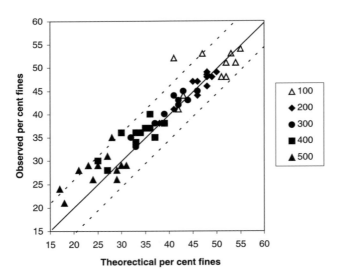

Figure 6.10 Concrete Series 1 – comparison between theoretical and observed per cent fines for fresh concrete of 50 mm slump made with Portland cements and a range of aggregates of 20 mm maximum size. Dotted lines are drawn at 6% on either side of the equality line.

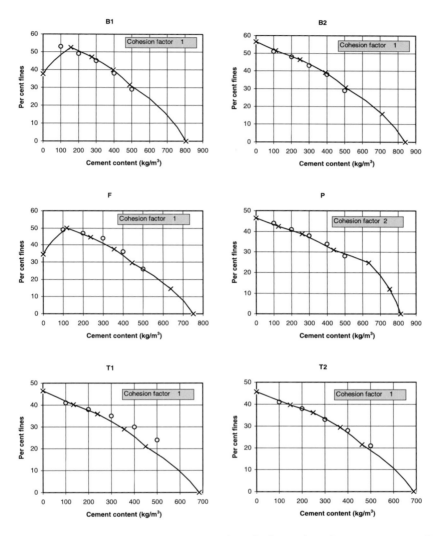

Figure 6.11 Concrete Series 1 – examples of relationships between per cent fines and cement content for concrete of 50 mm slump showing comparison between theory and observation.

The examples in Figure 6.11 show generally close agreement between theory and assessments by experienced supervisors but include particular cases of deviation. The diagram for B1 is an example where the per cent fines could have been reduced rather than increased for the leanest concrete. T1 is an example of increasing conservatism at higher cement contents leading to

Table 6.11 Concrete Series 1 – Predicted and observed fresh concrete densities

Materials code	Fresh concrete density (kg/m³)									
	Cement content (kg/m³)									
	100		200		300		400		500	
	Theory	Observed	Theory	Observed	Theory	Observed	Theory	Observed	Theory	Observed
B1	2248	2275	2319	2315	2358	2350	2376	2375	2371	2375
B2	2316	2316	2374	2383	2411	2421	2431	2449	2430	2404
F1	2238	2240	2298	2290	2330	2325	2342	2350	2331	2340
F2	2220	2200	2270	2268	2301	2304	2315	2320	2311	2310
G	2314	2336	2370	2372	2401	2395	2412	2404	2397	2404
H	2223	2200	2275	2280	2309	2330	2325	2330	2320	2325
I	2204	2180	2291	2260	2332	2320	2346	2340	2342	2340
K	2268	2260	2321	2315	2354	2365	2370	2375	2364	2375
N	2242	2240	2290	2285	2313	2310	2320	2320	2308	2315
P	2302	2320	2366	2360	2406	2400	2427	2430	2429	2430
T1	2279	2275	2349	2325	2380	2360	2390	2390	2373	2400
T2	2264	2280	2337	2350	2365	2370	2371	2360	2350	2340
Average	2260	2260	2322	2317	2355	2354	2369	2370	2361	2363

The entrapped air content was assumed to be 1% except for F1 and P for which 0.5% was assumed to match observed densities and water contents.

unnecessary over-cohesion. Example P is a case where additional conservatism applied throughout the range, leading to a cohesion factor of 2 compared with 1 for most of the materials combinations. This may be a case where the supervisor was aware that a commercial benefit could be exploited because of the relative prices of materials locally without having an adverse effect on concrete quality. Other possible reasons for additional conservatism could be either allowance for expected higher variation in materials qualities, or allowance for the expectancy of the local market with regard to the acceptable appearance or behaviour of the fresh concrete in practice.

Fresh concrete densities are listed in Table 6.11. Adjustments were made to the densities of one or both aggregates for five of the 12 combinations of materials as shown in Table 6.8.[8]

There is good agreement between the predicted densities (after minor adjustment) and observed densities of fresh concrete both on average and for the individual pairs of values.

Of 60 pairs of values for density, 93% are within ± 25 kg/m^3 and 73% are within ± 12 kg/m^3 of the line of equality in Figure 6.12.

Figure 6.13 shows two examples of very good agreement between theory and observation for relationships between concrete density and cement content when no adjustments were made to air content or materials densities.

The levelling of density and turndown at high cement contents are the expected results of the shape of the relationship between water demand and cement content.

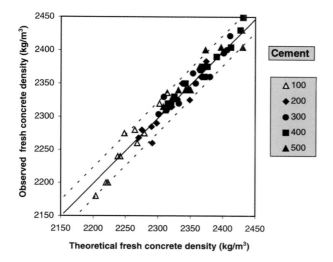

Figure 6.12 Concrete Series 1 – comparison between theoretical (adjusted) and observed densities of fresh concrete of 50 mm slump made with Portland cements and a range of aggregates of 20 mm maximum size. Dotted lines are drawn at 25 kg/m^3 on either side of the equality line.

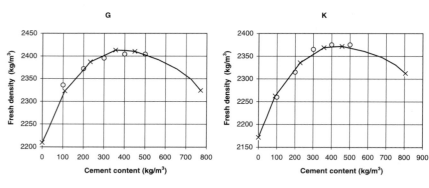

Figure 6.13 Concrete Series 1 – examples of relationships between density and cement content for fresh concrete of 50 mm slump showing comparison between theory and observation.

6.3.1.1 Overview of the results from Concrete Series 1

The close agreement between theory and observation provides evidence for the validity of extending the theory of particle mixtures to the prediction of relationships for concrete between free water demand, per cent fines, fresh concrete density and cement content under the following conditions:

Portland cement to BS 12
Natural gravel and crushed rock coarse aggregate of 20 mm max. size
Natural sand and mixtures with crushed rock fines.
No admixture
50 mm slump concrete
Cohesion factor 1 to 2
0.5 to 1% entrapped air

As a consequence of the goodness of fit for the investigated properties, the theory is deemed applicable also for derived values such as batch quantities of each material and commonly used ratios e.g. water/cement ratio and aggregate/cement ratio.

Evidence may be seen from Figure 6.14 for Concrete Series 1 of the ability of the theory to discriminate between materials to enable decisions to be made concerning the economy and effectiveness of different materials combinations.

It is clear from examination of Figure 6.14 that the ranking of materials with respect to water demand varies with cement content and that, at most cement contents over the full range, the theory would rank the materials in similar sequences with similar values of water demand to those obtained by concrete laboratory trials. As a result, valid comparisons could be made faster

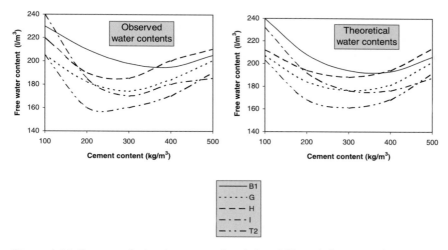

Figure 6.14 Concrete Series 1 – example of the ability of theory to discriminate between five sets of materials with respect to water content of concrete over the practical range of cement content.

and more cheaply by use of the theory, based on valid materials data, rather than by time consuming, tedious and expensive concrete laboratory trials.

6.3.2 Concrete Series 2 – including admixtures and air entrainment

The author accepted an offer from Mr P. Barnes, Technical Manager of Readicrete Limited, a member of the Readymix Group in the UK, to provide laboratory trial data on concrete for four sets of materials coded P1, P3, P4 and P5 for each of which, trials were made as follows

	Admixture
R1	None
R2	Plasticiser
R3	Air entrainer
R4	Plasticiser and air entrainer

The materials data and concrete trial data were analysed using the Theory of Particle Mixtures as extended to apply to concrete to determine the various additional factors necessary when particular admixtures are incorporated in the concrete. In all, 92 trial concrete batches were made at 50 mm slump. Measurements of total air content were made on concretes containing the air-entraining agent, otherwise air contents were assumed or estimated.

The cements were all Portland cements complying with BS 12 and the aggregates were 20–5 mm graded, predominantly flint pit gravels and natural siliceous pit sands complying with BS 882. The selected properties of the cements and aggregates together with adjusted values made as a result of analysis are summarised in Table 6.12.

The computer simulated values and observed results for water demand, percent fines, total air content, fresh concrete density and compressive strength are shown in Tables 6.13–6.28.

The values for the various factors for computer simulation, assessed from the results for water demand, per cent fines, total air content and fresh concrete density are summarised in Table 6.29. These factors are explained and discussed further in sections 5.1 and 5.2.

For concrete without admixtures, the average cohesion factor of 2.6 was higher than the normal value of unity and the average estimated entrapped air, 0.38%, was less than the normal value of 0.5%, but both were quite consistent for the tests of concretes made with the four different sets of materials at the one central laboratory. No significance for the theory is attached to the differences.

For the plasticised concretes, the voids factor reduced from unity to 0.88 implying a substantial reduction in water demand, however part of this reduction was accompanied by air entrainment for three of the four materials combinations. This is quite normal for the agent used, according to the manufacturer, and is dependent on interaction with the cement and, possibly more particularly, the fine aggregate. The cohesion factors reduced slightly as a result of the lower water demand and also the air-entrainment.

Table 6.12 Concrete Series 2 – Properties of materials including adjusted values

Data code	Materials	Properties				
		Mean size (mm)		Void ratio		Rel Densy ssd
		D	Adj	U	Adj	RD
P1	Cement	0.013		0.955	0.9	3.2
	Fine aggregate	0.595		0.66	0.6	2.60
	Coarse aggregate	10.71		0.8		2.55
P3	Cement	0.013		0.87	0.81	3.2
	Fine aggregate	0.61		0.67	0.58	2.62
	Coarse aggregate	10.69		0.66		2.56
P4	Cement	0.014		0.87		3.2
	Fine aggregate	0.65	0.53	0.71	0.60	2.61
	Coarse aggregate	9.63	10.5	0.79	0.74	2.57
P5	Cement	0.013		0.86		3.2
	Fine aggregate	0.53		0.58	0.61	2.65
	Coarse aggregate	11.4		0.77		2.55

Table 6.13 Concrete Series 2 – Summary of simulated and measured properties for materials combination P1, R1, without admixtures, at 50 mm slump

Concrete Series 2		P1 R1 No admixtures							
Comparison between simulated and trial data		Simul'n	Trial data	Simul'n	Trial data	Simul'n	Trial data	Simul'n	Trial data
Cement	kg/m³	120	120	209	209	272	272	324	324
Free water	kg/m³	217	207	193	190	180	188	179	178
Fine agg ssd	kg/m³	1024	1004	953	940	897	878	832	833
Coarse ssd	kg/m³	885	940	944	971	980	985	1007	1008
% fines		54	52	50	49	48	47	45	45
Plastic density	kg/m³	2246	2270	2299	2310	2330	2323	2342	2343
Total air	%	0.5		0.5		0.5		0.5	
28 day strength	N/mm²	3	4	19	19	35	35	45	47
Cement	kg/m³	372	372	427	427	473	473	524	524
Free water	kg/m³	178	181	179	171	183	184	189	191
Fine agg ssd	kg/m³	770	769	691	707	615	619	525	523
Coarse ssd	kg/m³	1033	1022	1064	1058	1091	1077	1122	1117
% fines		43	43	39	40	36	36	32	32
Plastic density	kg/m³	2352	2344	2360	2363	2361	2354	2361	2353
Total air	%	0.5		0.5		0.5		0.5	
28 day strength	N/mm²	54	56	62	65	66	65	68	68

Table 6.14 Concrete Series 2 – Summary of simulated and measured properties of concrete for materials combination P1 R2, incorporating a plasticiser, at 50 mm slump

Concrete Series 2		P1 R2 Plasticiser											
Comparison between simulated and trial data		Simul'n	Trial data	Simul'n	Trial data	Simul'n	Trial data	Simul'n	Trial data	Simul'n	Trial data	Simul'n	Trial data
Cement	kg/m³	206	206	272	272	323	323	370	370	421	421	469	469
Free water	kg/m³	163	174	157	159	158	155	157	153	159	154	163	164
Fine agg ssd	kg/m³	982	950	909	899	844	850	783	783	711	715	634	629
Coarse ssd	kg/m³	962	980	995	1007	1018	1018	1042	1042	1068	1069	1092	1097
% fines		51	49	48	47	45	45	43	43	40	40	37	36
Plastic density	kg/m³	2313	2310	2334	2338	2342	2345	2351	2348	2358	2360	2358	2358
Total air	%	1.8		1.8		1.8		1.8		1.8		1.8	
28 day strength	N/mm²	25	20.5	41	40	52	53	60	63	67	70	70	68

Table 6.15 Concrete Series 2 – Summary of simulated and measured properties of concrete for materials combination P1 R3, incorporating an air-entrainer, at 50 mm slump

Concrete Series 2		*P1 R3 Air-entrainer*											
Comparison between simulated and trial data		*Simul'n*	*Trial data*	*Simul'n*	*Trial data*	*Simul'n*	*Trial data*	*Simul'n*	*Trial data*	*Simul'n*	*Trial data*	*Simul'n*	*Trial data*
Cement	kg/m³	210	210	274	274	323	323	372	372	420	420	472	472
Free water	kg/m³	163	160	157	163	159	159	160	165	164	168	172	171
Fine agg ssd	kg/m³	913	889	848	829	783	776	718	716	646	645	558	569
Coarse ssd	kg/m³	957	979	985	993	1005	1006	1026	1018	1047	1039	1073	1072
% fines		49	48	46	45	44	44	41	41	38	38	34	35
Plastic density	kg/m³	2243	2240	2264	2260	2270	2265	2276	2270	2278	2271	2275	2286
Total air	%	4.5	5.6	4.5	4.7	4.5	5.2	4.5	4.8	4.5	4.8	4.5	4.7
28 day strength	N/mm²	21	18	35	33	42	42	49	50	54	53	57	59

Table 6.16 Concrete Series 2 – Summary of simulated and measured properties of concrete for materials combination P1 R4, incorporating an air-entrainer and plasticiser, at 50 mm slump

Concrete Series 2		*P1 R4 Air-entrainer and plasticiser*											
Comparison between simulated and trial data		*Simul'n*	*Trial data*	*Simul'n*	*Trial data*	*Simul'n*	*Trial data*	*Simul'n*	*Trial data*	*Simul'n*	*Trial data*	*Simul'n*	*Trial data*
Cement	kg/m³	207	207	275	275	322	322	369	369	422	422	469	469
Free water	kg/m³	149	167	145	149	146	152	147	148	151	138	158	151
Fine agg ssd	kg/m³	927	890	851	846	790	785	728	722	649	658	571	575
Coarse ssd	kg/m³	968	980	1000	1002	1019	1016	1040	1028	1064	1061	1086	1083
% fines		49	48	46	46	44	44	41	41	38	38	34	35
Plastic density	kg/m³	2252	2243	2270	2283	2277	2273	2284	2268	2286	2280	2284	2280
Total air	%	5	4.5	5	4.8	5	4.8	5	5.3	5	4.7	5	4.9
28 day strength	N/mm²	26	21	40	42	48	49.5	55	56	60	61	62	61

Table 6.17 Concrete Series 2 – Summary of simulated and measured properties of concrete for materials combination P3 R1, without admixtures, at 50 mm slump

Concrete Series 2		*P3 R1 No admixture*											
Comparison between simulated and trial data		*Simul'n*	*Trial data*	*Simul'n*	*Trial data*	*Simul'n*	*Trial data*	*Simul'n*	*Trial data*	*Simul'n*	*Trial data*	*Simul'n*	*Trial data*
Cement	kg/m³	124	124	214	214	274	274	314	314	432	432	546	546
Free water	kg/m³	195	188	175	171	165	168	163	160	162	165	176	174
Fine agg ssd	kg/m³	987	931	909	874	844	822	791	781	616	657	419	454
Coarse ssd	kg/m³	984	1065	1040	1096	1082	1111	1106	1125	1185	1146	1250	1230
% fines		50	47	47	44	44	43	42	41	34	36	25	27
Plastic density	kg/m³	2291	2298	2338	2339	2365	2372	2374	2374	2395	2400	2391	2405
Total air	%	0.5		0.5		0.5		0.5		0.5		0.5	
28 day strength	N/mm²	3	4.5	23	20	39	40	49	50	68	68	73	74

Table 6.18 Concrete Series 2 – Summary of simulated and measured properties of concrete for materials combination P3 R2, incorporating a plasticiser, at 50 mm slump

Concrete Series 2		*P3 R2 Plasticiser*									
Comparison between simulated and trial data		*Simul'n*	*Trial data*	*Simul'n*	*Trial data*	*Simul'n*	*Trial data*	*Simul'n*	*Trial data*	*Simul'n*	*Trial data*
Cement	kg/m³	123	123	216	216	318	318	445	445	550	550
Free water	kg/m³	185	196	162	166	153	151	153	151	165	166
Fine agg ssd	kg/m³	1010	934	919	872	790	786	597	628	426	454
Coarse ssd	kg/m³	989	1067	1061	1096	1129	1135	1217	1185	1268	1239
% fines		51	47	46	44	41	41	33	35	25	27
Plastic density	kg/m³	2307	2321	2358	2350	2390	2379	2412	2409	2409	2409
Total air	%	0.5		0.5		0.5		0.5		0.5	
28 day strength	N/mm²	4	4.5	27	30	54	56	71	72	75	73

Table 6.19 Concrete Series 2 – Summary of simulated and measured properties of concrete for materials combination P3 R3, incorporating an air-entrainer, at 50 mm slump

Concrete Series 2		P3 R3 Air-entrainer									
Comparison between simulated and trial data		*Simul'n*	*Trial data*	*Simul'n*	*Trial data*	*Simul'n*	*Trial data*	*Simul'n*	*Trial data*	*Simul'n*	*Trial data*
Cement	kg/m³	121	121	210	210	305	305	423	423	532	532
Free water	kg/m³	157	156	146	138	144	138	152	153	173	164
Fine agg ssd	kg/m³	944	889	846	823	724	731	535	570	359	400
Coarse ssd	kg/m³	1030	1082	1080	1100	1131	1126	1199	1159	1232	1225
% fines		48	45	44	43	39	39	31	33	23	25
Plastic density	kg/m³	2252	2249	2283	2270	2303	2300	2309	2305	2295	2320
Total air	%	4.3	4.4	4.3	5.2	4.3	4.0	4.3	4.0	4.3	3.7
28 day strength	N/mm²	6	3.5	26	20	46	48	59	58	63	66

Table 6.20 Concrete Series 2 – Summary of simulated and measured properties of concrete for materials combination P3 R4, incorporating an air-entrainer and plasticiser, at 50 mm slump

Concrete Series 2			P3 R4 Air-entrainer and plasticiser							
Comparison between simulated and trial data		Simul'n	Trial data	Simul'n	Trial data	Simul'n	Trial data	Simul'n	Trial data	Simul'n	Trial data
Cement	kg/m^3	120	120	213	213	310	310	428	428	535	535
Free water	kg/m^3	148	165	136	138	135	131	144	145	163	164
Fine agg ssd	kg/m^3	951	890	848	823	718	737	529	576	365	386
Coarse ssd	kg/m^3	1027	1080	1084	1105	1136	1140	1204	1171	1231	1236
% fines		48	45	44	43	39	39	31	33	23	24
Plastic density	kg/m^3	2246	2256	2281	2278	2299	2295	2305	2320	2293	2320
Total air	%	5	4.7	5	5.8	5	5.5	5	5.4	5	5.3
28 day strength	N/mm^2	6	5	27	27	46	50	57	57	61	60

Table 6.21 Concrete Series 2 – Summary of simulated and measured properties of concrete for materials combination P4 R1, without admixtures, at 50 mm slump

Concrete Series 2		P4 R1 No admixtures								
Comparison between simulated and trial data		Simul'n	Trial data	Simul'n	Trial data	Simul'n	Trial data	Simul'n	Trial data	Simul'n	Trial data
Cement	kg/m³	122	122	214	214	309	309	436	436	541	541
Free water	kg/m³	214	199	189	179	174	176	172	179	181	181
Fine agg ssd	kg/m³	1019	985	948	932	857	862	694	706	525	545
Coarse ssd	kg/m³	913	988	972	1009	1023	1013	1089	1058	1148	1123
% fines		53	50	49	48	46	46	39	40	31	33
Plastic density	kg/m³	2268	2294	2324	2334	2364	2360	2391	2379	2394	2390
Total air	%	0.25		0.25		0.25		0.25		0.25	
28 day strength	N/mm²	3	5	20	22	42	42	63	62	68	65

Table 6.22 Concrete Series 2 – Summary of simulated and measured properties of concrete for materials combination P4 R2, incorporating a plasticiser, at 50 mm slump

Concrete Series 2		*P4 R2 Plasticiser*									
Comparison between simulated and trial data		*Simul'n*	*Trial data*	*Simul'n*	*Trial data*	*Simul'n*	*Trial data*	*Simul'n*	*Trial data*	*Simul'n*	*Trial data*
Cement	kg/m³	120	120	209	209	311	311	437	437	542	542
Free water	kg/m³	193	197	180	180	168	169	166	164	174	179
Fine agg ssd	kg/m³	1050	1003	969	957	865	869	706	715	540	552
Coarse ssd	kg/m³	925	965	966	978	1016	1016	1079	1068	1135	1124
% fines		53	51	50	49	46	46	40	40	32	33
Plastic density	kg/m³	2288	2285	2324	2324	2361	2348	2387	2384	2392	2396
Total air	%	0.75		0.75		0.75		0.75		0.75	
28 day strength	N/mm²	4	5	22	23	47	50	69	66	73	73

Table 6.23 Concrete Series 2 – Summary of simulated and measured properties of concrete for materials combination P4 R3, incorporating an air-entrainer, at 50 mm slump

Concrete Series 2											
Comparison between simulated and trial data		P4 R3 Air-entrainer									
		Simul'n	Trial data	Simul'n	Trial data	Simul'n	Trial data	Simul'n	Trial data	Simul'n	Trial data
Cement	kg/m³	124	124	208	208	301	301	418	418	522	522
Free water	kg/m³	173	175	160	149	153	152	157	162	170	172
Fine agg ssd	kg/m³	987	978	912	898	816	817	663	657	497	483
Coarse ssd	kg/m³	937	969	978	986	1016	1019	1063	1053	1109	1113
% fines		51	50	48	48	45	45	38	38	31	30
Plastic density	kg/m³	2222	2245	2258	2240	2286	2290	2301	2290	2298	2290
Total air	%	4.5	4.0	4.5	5.6	4.5	4.9	4.5	4.4	4.5	3.8
28 day strength	N/mm²	4	5.5	18	18	35	39	48	47	53	51

Table 6.24 Concrete Series 2 – Summary of simulated and measured properties of concrete for materials combination P4 R4, incorporating an air-entrainer and plasticiser, at 50 mm slump

Concrete Series 2		P4 R4 Air-entrainer and plasticiser								
Comparison between simulated and trial data		Simul'n	Trial data	Simul'n	Trial data	Simul'n	Trial data	Simul'n	Trial data	Simul'n	Trial data
Cement	g/m^3	123	123	208	208	306	306	423	423	538	538
Free water	g/m^3	158	166	147	143	143	142	146	156	161	160
Fine agg ssd	g/m^3	996	982	915	912	806	825	652	667	469	500
Coarse ssd	g/m^3	959	975	997	997	1038	1033	1086	1069	1137	1106
% fines		51	50	48	48	44	44	38	38	29	31
Plastic density	g/m^3	2236	2245	2268	2260	2293	2305	2308	2310	2305	2311
Total air	%	4.9	4.8	4.9	5.2	4.9	5.2	4.9	5	4.9	4.4
28 day strength	N/mm^2	7	7.5	25	27	44	46.5	56	55	61	58

Table 6.25 Concrete Series 2 – Summary of simulated and measured properties of concrete for materials combination P5 R1, without admixtures, at 50 mm slump

Concrete Series 2		P5 R1 No admixture									
Comparison between simulated and trial data		Simul'n	Trial data	Simul'n	Trial data	Simul'n	Trial data	Simul'n	Trial data	Simul'n	Trial data
Cement	kg/m³	121	121	213	213	307	307	441	441	552	552
Free water	kg/m³	206	210	182	179	170	176	171	167	187	178
Fine agg ssd	kg/m³	971	948	900	873	806	794	624	652	440	449
Coarse ssd	kg/m³	988	1036	1045	1074	1091	1092	1156	1136	1203	1217
% fines		50	48	46	45	42	42	35	36	27	27
Plastic density	kg/m³	2286	2315	2339	2339	2373	2370	2392	2396	2382	2396
Total air	%	0.25		0.25		0.25		0.25		0.25	
28 day strength	N/mm²	4	5	23	24	46	45.5	66	66	70	70

Table 6.26 Concrete Series 2 – Summary of simulated and measured properties of concrete for materials combination P5 R2, incorporating a plasticiser, at 50 mm slump

Concrete Series 2		P5 R2 Plasticiser									
Comparison between simulated and trial data		Simul'n	Trial data	Simul'n	Trial data	Simul'n	Trial data	Simul'n	Trial data	Simul'n	Trial data
Cement	kg/m^3	124	124	215	215	312	312	444	444	543	543
Free water	kg/m^3	160	180	151	164	150	146	156	136	169	164
Fine agg ssd	kg/m^3	1010	955	918	881	802	809	620	660	464	455
Coarse ssd	kg/m^3	1028	1052	1065	1090	1102	1117	1158	1163	1196	1232
% fines		50	48	46	45	42	42	35	36	28	27
Plastic density	kg/m^3	2321	2310	2349	2340	2366	2360	2377	2395	2372	2395
Total air		1.75		1.75		1.75		1.75		1.75	
28 day strength	N/mm^2	7	7	30	26	51	52	67	65	70	71

Table 6.27 Concrete Series 2 – Summary of simulated and measured properties of concrete for materials combination P5 R3, incorporating an air-entrainer, at 50 mm slump

Concrete Series 2		P5 R3 Air-entrainer									
Comparison between simulated and trial data		Simul'n	Trial data	Simul'n	Trial data	Simul'n	Trial data	Simul'n	Trial data	Simul'n	Trial data
Cement	kg/m³	117	117	208	208	296	296	426	426	527	527
Free Water	kg/m³	166	186	152	151	149	144	159	162	177	177
Fine agg ssd	kg/m³	937	902	855	824	760	749	580	625	416	457
Coarse ssd	kg/m³	999	1041	1040	1077	1070	1087	1114	1128	1145	1185
% fines		48	46	45	43	42	41	34	36	27	28
Plastic density	kg/m³	2219	2245	2256	2240	2275	2255	2279	2340	2265	2345
Total air	%	5.2	3.7	5.2	6.2	5.2	6.5	5.2	5.6	5.2	4
28 day strength	N/mm²	5	5	24	25	40	37	55	57	59	58

Table 6.28 Concrete Series 2 – Summary of simulated and measured properties of concrete for materials combination P5 R4, incorporating an air-entrainer and plasticiser, at 50 mm slump

Concrete Series 2		P5 R4 Air-entrainer and plasticiser									
Comparison between simulated and trial data		*Simul'n*	*Trial data*	*Simul'n*	*Trial data*	*Simul'n*	*Trial data*	*Simul'n*	*Trial data*	*Simul'n*	*Trial data*
Cement	kg/m³	119	119	206	206	306	306	426	426	524	524
Free Water	kg/m³	146	156	136	137	136	136	144	143	159	157
Fine agg ssd	kg/m³	959	899	875	835	757	757	592	610	440	438
Coarse ssd	kg/m³	1019	1047	1056	1089	1091	1112	1135	1131	1164	1196
% fines		48	46	45	43	41	41	34	35	27	27
Plastic density	kg/m³	2243	2175	2273	2255	2290	2310	2296	2305	2286	2305
Total air	%	5.5	5.7	5.5	5.9	5.5	5	5.5	5.7	5.5	5.1
28 day strength	N/mm²	6	7	25	24	43	47	55	56	59	59

Table 6.29 Concrete Series 2 – Computer Simulation Factors for fresh concrete estimated from the data to account for the use of admixtures

Data code	Admixture type	Factors	Air factors			Cohesion factors		Air content (%) m = measured		
		Admixture voids factor	a	b	c	Overall incl. air	Air	Entrapped	Entrained	Total
P1 R1		1				2.5	0.00	0.5	0	0.5
P3 R1		1				2.5	0.00	0.5	0	0.5
P4 R1		1				3.25	0.00	0.25	0	0.25
P5 R1		1				2	0.00	0.25	0	0.25
Average		1	53	135	0.6	2.6	0	0.38		0.38
P1 R2	Plas	0.85	53	135	0.6	2	−0.26	0.5	1.3	1.8
P3 R2	Plas	0.87				2	0.00	0.5	0	0.5
P4 R2	Plas	0.95	53	135	0.6	3.25	−0.10	0.25	0.5	0.75
P5 R2	Plas	0.85	53	135	0.6	1.5	−0.30	0.25	1.50	1.75
Average		0.88	53	135	0.6	2.2	−0.17	0.38	0.83	1.20
P1 R3	AEA	1	53	135	0.6	2	−0.80	0.5	4	4.5(m)
P3 R3	AEA	1	53	135	0.6	1.5	−0.76	0.5	3.8	4.3(m)
P4 R3	AEA	1	53	135	0.6	2.75	−0.85	0.25	4.25	4.5(m)
P5 R3	AEA	1	53	135	0.6	1.5	−0.99	0.25	4.95	5.2(m)
Average		1.00	53	135	0.6	1.9	−0.85	0.38	4.25	4.63
P1 R4	Plas/AEA	0.87	53	135	0.6	1.5	−0.90	0.5	4.5	5.0(m)
P3 R4	Plas/AEA	0.92	53	135	0.6	1.25	−0.90	0.5	4.5	5.0(m)
P4 R4	Plas/AEA	0.9	53	135	0.6	2.25	−0.93	0.25	4.65	4.9(m)
P5 R4	Plas/AEA	0.85	53	135	0.6	1	−1.05	0.25	5.25	5.5(m)
Average		0.89	53	135	0.6	1.5	−0.95	0.38	4.73	5.10

Table 6.30 Concrete Series 2 – Factors for compressive strength of concrete estimated from the data to account for the use of admixtures

| Data code | Admixture | Strength factors | | | | | | Air | |
		Intercept (Cem't str)	Slope (Aggreg)	Age 3d	4d	7d	28d (assumed)	Entrapped (assumed)	Entrained (estimated)
P1 R1	None	52	1.01				1	−0.038	
R2	Plas	58	1.08				1	−0.038	−0.015
R3	AEA	58	1.09				1	−0.038	−0.025
R4	Plas/AEA	59	1.04				1	−0.038	−0.025
	Ave	57	1.06				1	−0.038	
P3 R1	None	58	1.11	1.35		1.24	1	−0.038	
R2	Plas	57	1.09		1.38	1.15	1	−0.038	
R3	AEA	60	1.11	1.49		1.23	1	−0.038	−0.025
R4	Plas/AEA	57	1.1	1.50		1.20	1	−0.038	−0.025
	Ave	58	1.10	1.45	1.32	1.21	1	−0.038	
P4 R1	None	51	1.05	1.48		1.26	1	−0.038	
R2	Plas	54	1.03	1.44		1.21	1	−0.038	−0.015
R3	AEA	50	1.10	1.41		1.23	1	−0.038	−0.025
R4	Plas/AEA	52	1.00		1.32	1.16	1	−0.038	−0.025
	Ave	52	1.05	1.44		1.22	1	−0.038	
P5 R1	None	51	1.00	1.33		1.14	1	−0.038	
R2	Plas	54	1.06	1.34		1.15	1	−0.038	−0.015
R3	AEA	56	1.04	1.35		1.20	1	−0.038	−0.025
R4	Plas/AEA	56	1.10	1.32		1.15	1	−0.038	−0.025
	Ave	54	1.05	1.34		1.16	1	−0.038	

For the air-entrained concretes, an assumption was made of no direct water reduction, and the entrapped air contents were taken to be the same as for the corresponding non-air-entrained concretes. The entrained air was calculated as the measured total air less the assumed entrapped air. On average 4.25% entrained air was estimated to have been generated by use of the agent. The cohesion factor was reduced by the presence of the entrained air.

For the plasticised air-entrained concretes, the plasticising effect was estimated to be very similar to that obtained for the plasticiser alone, as judged by the similarity of the voids factors. The resulting air entrainment was estimated to be greater than that associated with the air-entraining agent acting on its own, but not quite as great as combining the separate effects of the air-entraining agent and plasticiser. The factors for cohesion were reduced as a result of the combined effect.

The single set of values for the entrained-air factors, a, b and c, which were used to relate water reduction and cement content, did not require to be modified for any particular admixture or materials combination.

Thus, the data and inferred values for the various factors are mutually supportive and provide confidence that the theory has taken the effects of admixtures satisfactorily into account.

The values for the various factors would not necessarily be valid for other plasticisers, air-entraining agents or other combinations of materials. However they would be a useful basis for assessing other combinations.

The values for the various factors discussed in section 5.4 were assessed from the results for strength and are summarised in Table 6.30. Results concerning each of the four materials combinations have been kept together because of the predominant effects of cement and aggregate on the values of the factors. No obvious or consistent effects of admixtures were detected, except with regard to the factor for air in the final two columns. These show lower values for the index for entrained air than for entrapped air and a further reduction when low values of entrained air were involved. Otherwise, the same factors can reasonably be assumed to apply to concrete with and without plasticising and/or air-entraining admixtures.

Examples from Concrete Series 2 of the various important relationships are shown in Figures 6.15–6.18 for concretes for materials combination P1, with and without admixtures. These diagrams illustrate good agreement between theory and laboratory data and assist in confirming the validity of the extension of the theory to cover concretes with air-entraining and plasticising admixtures and to cover strength.

Summarised comparisons are provided in Figure 6.19 between theoretical and observed properties of concrete in Series 2, for concretes with and without admixtures. Arbitrary limit lines are shown on either side of the equality lines to aid comparisons.

For water demand, 79 of 92 pairs of values are within $\pm 10 \ \text{l/m}^3$. No systematic trends are in evidence.

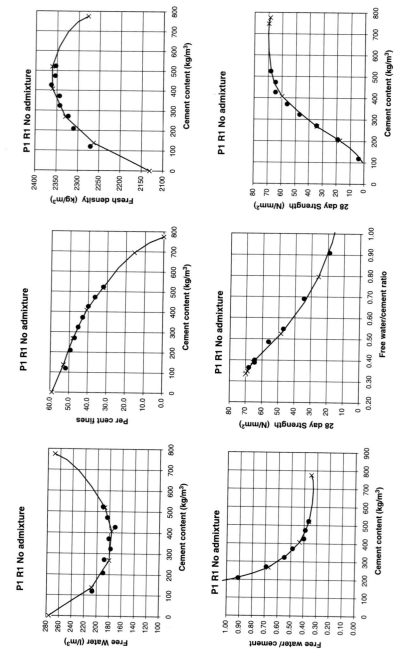

Figure 6.15 Concrete Series 2 – relationships for concrete containing materials coded P1 R1 without admixtures.

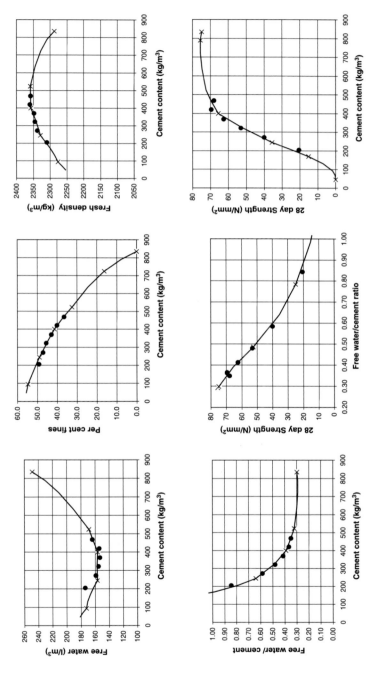

Figure 6.16 Concrete Series 2 – relationships for concrete containing materials coded P1 R2 with a plasticiser.

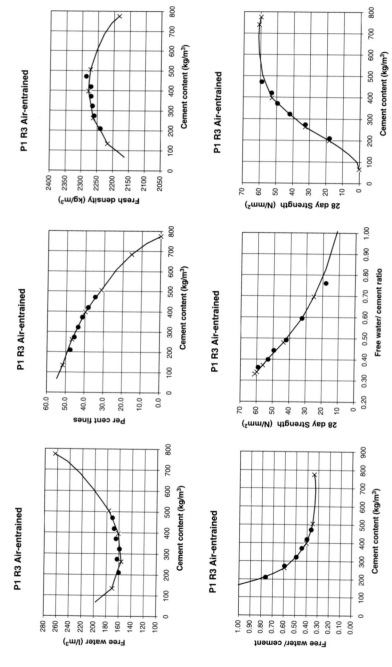

Figure 6.17 Concrete Series 2 – relationships for concrete containing materials coded P1 R3 with an air-entraining agent.

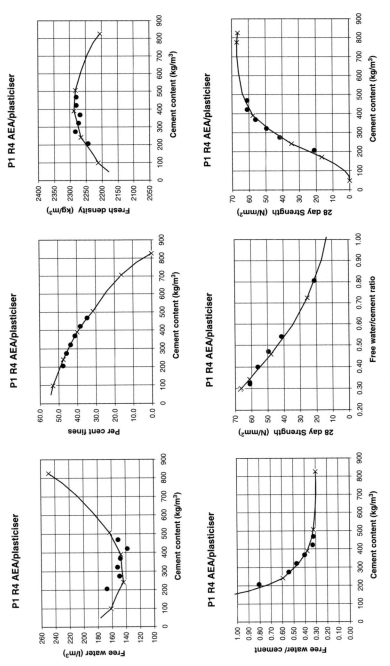

Figure 6.18 Concrete Series 2 – relationships for concrete containing materials coded P1 R4 with both plasticiser and air-entraining agent.

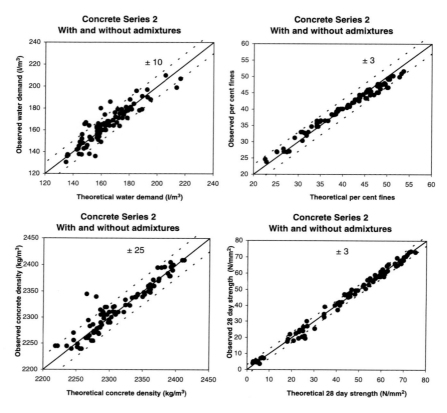

Figure 6.19 Concrete Series 2 – comparison between theoretical concrete properties for concrete with and without admixtures.

For per cent fines, 91 of 92 pairs of values are within ±3%. There is a
strong tendency for the observed values for Series 2 to be lower than the
theoretical values at low cement contents and the reverse at high cement
contents.

For density, 85 of 92 pairs of values are within ±25 kg/m³.

For strength, 80 of 92 pairs of values are within ±3 N/m³.

*Overview of the results from Concrete Series 2 for concrete with
admixtures and air entrainment*

The results confirm that the principles developed for concrete without
admixtures may be extended to concretes with plasticisers and air
entrainment by suitable modification to allow for water reduction and the
effects of air content and water reduction on strength.

Table 6.31 Concrete Series 3 – Summary of properties of materials for fly ash concrete

Materials code	Mean size (mm)				Void ratio				Density (kg/m³)			
	Cement	Addition	Fine agg	Coarse agg	Cement	Addition	Fine agg	Coarse agg	Cement	Addition	Fine agg ssd	Coarse agg ssd
N & N2	0.015	0.0105	0.73	11.1	0.87	0.52	0.55	0.605	3200	2325	2580	2440
adj					0.85		0.50					2490

Note
The fly-ash was pulverized fuel ash to BS 3892 Part 1.

Table 6.32 Computer simulation of combination of fly-ash and Portland cement

Combination of fly-ash and cement

Materials data	Proportion (%)	Propn by vol (%)	Mean size (mm)	Voids ratio	Relative density SSD
Fly-ash	25	31	0.0105	0.54	2.325
Portland cement	75	69	0.0150	0.85	3.2
Combination	100	100	0.0134	0.720	2.92

Table 6.33 Concrete Series 3 – Predicted and observed water demands at 50 mm slump for concretes with and without fly-ash

Materials code		Water content (l/m^3)									
		Cement + fly-ash content (kg/m^3)									
		100		200		300		400		500	
		Theory	Observed	Theory	Observed	Theory	Observed	Theory	Observed	Theory	Observed
N	Cement	188	190	168	175	164	167	169	168	188	180
N2	30% fly-ash 70% cement	184	170	159	158	162	150	157	158	176	168

Note
The air content was assumed to be 1% and the cohesion factor to be 1.5.

Table 6.34 Concrete Series 3 – Predicted and observed per cent fines at 50 mm slump for concretes with and without fly-ash

Materials code		Per cent fines									
		Cement + fly-ash content (kg/m³)									
		100		200		300		400		500	
		Theory	Observed	Theory	Observed	Theory	Observed	Theory	Observed	Theory	Observed
N	Cement	51	48	46	44	39	40	30	36	21	28
N2	30% fly-ash 70% cement	51	47	45	44	38	40	29	35	19	28

Note
The cohesion factor was assumed to be 1.5.

Table 6.35 Concrete Series 3 – Predicted and observed fresh densities at 50 mm slump for concretes with and without fly-ash

Materials code	Fresh concrete density (kg/m^3)									
	Cement + fly-ash content (kg/m^3)									
	100		200		300		400		500	
	Theory	*Observed*	*Theory*	*Observed*	*Theory*	*Observed*	*Theory*	*Observed*	*Theory*	*Observed*
N Cement	2242	2240	2290	2285	2313	2310	2320	2320	2308	2315
N2 30% fly-ash 70% cement	2241	2235	2288	2280	2307	2310	2308	2320	2288	2315

Note
The air content was assumed to be 1%.

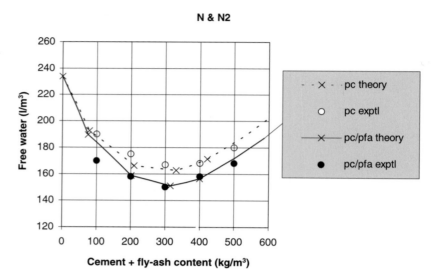

Figure 6.20 Concrete Series 3 – Comparison between computer simulated values and observed results for water demands for concretes with and without fly-ash.

Table 6.36 Concrete Series 3 – Predicted and observed 28 day strengths for concretes with and without ground granulated blastfurnace slag

Materials code

P1R1	without ggbs		P1R1G	with ggbs	
Free w/c	28d strength (N/mm^2)		Free $w/c + e \times addn$	28d strength (N/mm^2)	
	Theory	Observed		Theory	Observed
1.73	3	4			
0.91	19	19	0.94	18	19
0.69	35	35	0.72	31	34
0.55	45	47	0.57	44	45
0.49	54	56	0.50	51	52
0.40	62	65	0.42	63	61
0.39	66	65	0.40	64	63
0.36	68	68			

Note
P1R1G ggbs/pc = 40/60 by mass; $e = 0.91$.

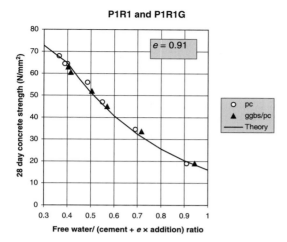

Figure 6.21 Relationships between 28 day strength of concrete and free water/ (cement + *e* × addition).

6.3.3 Concrete Series 3 – including additions

In this section are described the test results, and comparisons with theory, for concretes made with fly ash or ground granulated blastfurnace slag.

Concretes with fly-ash

One series of concrete mixtures was made with a 25/75% combination by mass of fly-ash/cement; the fly-ash used was pulverised-fuel ash complying with BS 3892 Part 1. The properties of the materials are summarised in Table 6.31.

For the computer simulation, the properties of the combination of fly-ash and cement were first calculated, as shown in Table 6.32, before combining with the fine and coarse aggregates.

The measured voids ratio of the combination using the Vicat test was 0.74 compared with the estimated value of 0.72 by computer simulation.

It will be observed in Table 6.32 that the proportion by volume of fly-ash in the combination is 31% by volume compared with 25% by mass, due to the difference in relative density between the two materials.

The concrete test results are summarised in Tables 6.33–6.35 for concrete with and without fly-ash.

The results for water demand for concretes with and without fly-ash are illustrated by Figure 6.20.

It will be observed from Figure 6.20 that the theoretical reduction in water demand associated with the use of fly-ash is generally confirmed by the experimental data although the predicted trend with cement content is not matched.

Concretes with ground granulated blastfurnace slag

Six of the eight batches of concrete reported under Concrete Series 1 P1R1 were remade as Series 3 P1R1G but with a combination of 40/60% of ground granulated blastfurnace slag to BS 6699 and Portland cement. Otherwise the mix proportions were the same as for P1R1 with Portland cement. Strengths at 28 days were measured for comparison with the results from P1R1 at the same water contents and same mass of total cementitious material. The value of the efficiency factor e was estimated to be 0.91 to enable the plotted results for the ggbs combinations to lie approximately on the same curve as for the Portland cement concrete when plotting strength against the ratio of water/(cement $+ e \times$ addition).

The results are shown in Table 6.36 and Figure 6.21.

The inclusion of the simple strength efficiency factor, $e = 0.91$ at 28 days in this instance, has permitted the relationship between strength and w/c to be extended to the use of ground granulated blastfurnace slag in combination with Portland cement.

Overview of the results from Concrete Series 3 for concrete with additions

The results provide useful confirmation of the validity of extending the theory to the use of additions such as fly-ash and ground granulated blastfurnace slag. Another addition is considered in section 8.1, as a special case study.

7 Proportions and properties of concrete predicted by the use of the theory of particle mixtures

The purpose of this section is to demonstrate the use of the Theory of Particle Mixtures as a tool for diagnosis, development and education.

7.1 Fine/total aggregate percentage (per cent fines)

The percentage of fine to total aggregate, (or per cent fines), is influenced by the cement content and by the mean size of fine aggregate as shown in Figure 7.1.

Hughes (1960) recognised that when a coarse sand is used there might be a point at which satisfactory cohesive concrete is not achievable unless cement content is increased or the sand changed. This is compatible with the situation in Figure 7.1 where, for the two coarsest fine aggregates, when the cement contents are less than 150 kg/m³, the optimum per cent fines decreases.

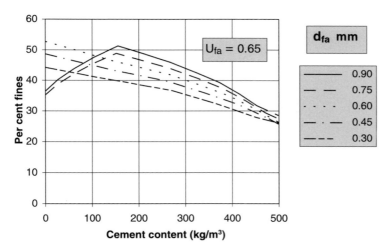

Figure 7.1 Influence of mean size of fine aggregate and cement content on the per cent fines.

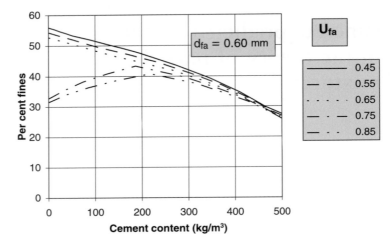

Figure 7.2 Influence of voids ratio of fine aggregate and cement content on the per cent fines.

Glanville *et al.* (1938) recognised that aggregate grading is less important for rich concretes and that low cement contents require finer grading, which implies either a higher sand content or the use of a finer sand, as illustrated by Figure 7.1.

Higher voids ratios of fine aggregates require lower per cent fines to avoid over-filling of voids in the coarse aggregate, as shown in Figure 7.2. For the

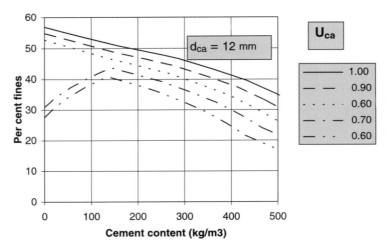

Figure 7.3 Influence of voids ratio of coarse aggregate and cement content on the per cent fines.

two highest void ratios, cement contents below about 200 kg/m³ require more substantial reductions in per cent fines. Thus, no improvement to cohesion can be expected for lean concretes by using a higher per cent fines when the fine aggregate is either very coarse, as in Figure 7.1, or has a high voidage as in Figure 7.2. Both of these accord with the author's practical experience, that with some materials there may be a cement content below which increasing the per cent fines is not beneficial for cohesion.

The effects of voids ratio of the coarse aggregate are shown in Table 7.3 from which it will be seen that increasing the voids ratio of the coarse aggregate requires increased per cent fines.

The combined effects of maximum aggregate size with the consequent effects of coarse aggregate voids ratio, are displayed later in section 7.2.2 and Figure 7.7.

7.2 Water demand

7.2.1 Influence of cement properties on concrete water demand

The aspects discussed here are supported in Appendix C by the detailed examination of data from the literature.

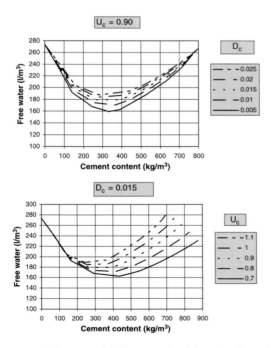

Figure 7.4 Theoretical influence of voids ratio U_c and mean size D_c of the cement on the water demand of concrete.

Figure 7.4, utilising the Theory of Particle Mixtures, demonstrates that a lower mean size of cement reduces the water demand throughout the practical range of cement content, and particularly at the centre of the range where particle interference will be the greatest. A lower void ratio of the cement also reduces the water demand of the concrete and the benefit increases with cement content, as might be expected.

Thus, with regard to water demand for concretes of high cement content, mean size of the cement is less important and voids ratio more important.

Combined theoretical effects of mean size and void ratio of cement on the water demand of concrete

In the manufacture of cement, a reduction in mean size resulting from finer grinding may be accompanied by a steeper particle size distribution, i.e. a smaller size range, and as a result the void ratio may be increased. Thus, the net effect of employing a finer cement is often a higher water demand for the more critical rich concretes. The composite effect of a reduction in size will vary, dependent on the extent to which the slope of the distribution is maintained by finer grinding, and the extent to which the fine end is modified particularly by any changes to the gypsum content of the cement. Examples of possible results of finer grinding are illustrated by Figure 7.5.

Thus, from Figure 7.5 may be seen the significance of changes in cement properties especially for the richer concretes. In particular, the substantial benefit is apparent when it is possible to reduce voids ratio and mean size together.

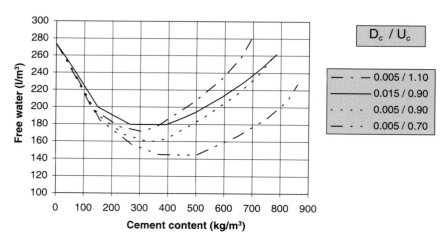

Figure 7.5 Examples of the combined effects on concrete due to finer grinding of a cement dependent on the extent to which the void ratio is increased, maintained or reduced.

7.2.2 *Influence of coarse aggregate on concrete water demand*

The theoretical effects of size and voids ratio of the coarse aggregate are illustrated in Figure 7.6. It should be noted that the quoted values of size are mean size **not** maximum size. Coarse aggregates having maximum sizes of 10 mm, 20 mm and 40 mm might have mean sizes of 7, 12 and 17 mm respectively, corresponding approximately to the middle three relationships in the upper diagram of Figure 7.6.

Generally, water demand reduces with increasing mean size and reducing void ratio of the coarse aggregate at all cement contents in the working range but the benefit is greater at the lower cement contents. This accords with Bloem and Gaynor (1963) who found a tendency for water demand of concrete to reduce as coarse aggregate void **content** reduced.

In practice, the use of a larger maximum size of aggregate permits a larger range of size fractions to be used, so that the water demand should reduce for two reasons. These are the reduced voids ratio of the coarse material and the reduction in particle interference, because of the reduced size ratio of mortar solids to coarse aggregate.

The possibility of some size fractions having poor shape, due to nature or to processing, needs to be taken into account when comparing different

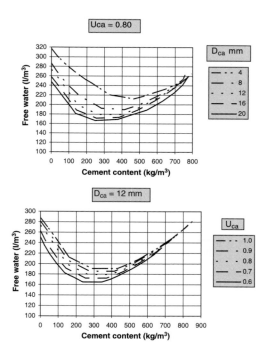

Figure 7.6 Theoretical influence of mean size and voids ratio of coarse aggregate on the water demand of concrete.

maximum sizes of apparently the same material. For example, in crushing oversize material to ensure sufficient 20 mm aggregate of regular shape, it often occurs that the 10 mm material is affected adversely, such that an increase in void ratio occurs for this size fraction; this is usually offset by reducing, or omitting altogether, the 10 mm fraction.

Combined theoretical effects of mean size and void ratio of coarse aggregate on the water demand of concrete

The consequent combined effects of changes in the maximum size of aggregate on concrete water demand are illustrated in Figure 7.7, for aggregates complying with BS 882 for graded coarse aggregates, and having the properties shown in Table 7.1.[1]

It will be seen from Table 7.1 that the voids ratios reduce significantly as the maximum aggregate size is increased.

The benefits for the water demand of a larger mean size of aggregate and the associated reduced voids ratio are apparent in Figure 7.7, the benefits being greatest for low and medium cement contents. Lower values of per cent fine/total aggregate are required as coarse aggregate size is increased.

The effects shown in Figure 7.7 are compatible with the literature. For example, Walker and Bloem (1960) demonstrated reductions in water demand of about 20 l/m^3 in changing the maximum size from 9.5 to

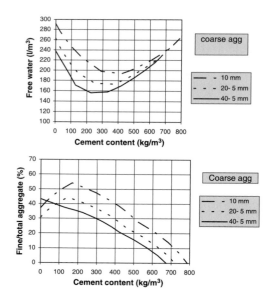

Figure 7.7 Theoretical influence of maximum aggregate size on water demand of concrete taking account of consequent changes in voids ratio of the coarse aggregate and changes to the per cent fines.

Table 7.1 Assumed properties of coarse aggregates

BS 882 designat'n	40- 5 mm	20- 5 mm	10 mm
Mean size (mm)	16.2	10.7	6.6
Voids ratio	0.60	0.67	0.77

19 mm, and a further reduction of about 15 l/m^3 in changing to 37.5 mm. Teychenne *et al.* (1988) also shows changes averaging 20 l/m^3 for similar changes in mean size.

7.2.3 *Influence of fine aggregate properties on concrete water demand*

The theoretical effects of mean size and voids ratio of the fine aggregate on water demand of concrete are illustrated in Figure 7.8. All concretes have been optimised for adequate cohesion.

It may be seen from Figure 7.8, that water demand reduces with reducing mean size of the fine aggregate to a generally similar extent over the practical

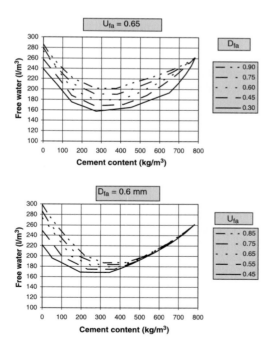

Figure 7.8 Theoretical influence of mean size and voids ratio of fine aggregate on the water demand of concrete.

working range of cement content, whereas the benefit of lower voids ratios in reducing water demand decreases with cement content.

Wills (1967) found that a change in fine aggregate shape as represented by change in voids **content** produced a change in water demand double or treble that associated with the same change in coarse aggregate shape. Comparison of Figures 7.6 and 7.8 indicates partial agreement with this observation, in that for the particular mean sizes selected for the lower diagrams, a reduction in voids ratio of the fine aggregate resulted in a reduction in water demand of about twice that for the same reduction in voids ratio for the coarse aggregate at very low cement contents, but the same reduction at medium cement contents.

A smaller mean size of fine aggregate should result in a benefit throughout the range of cement content, but a low voids ratio is also important at low cement contents. However, in practice, the smaller sized fine aggregates often have higher voids ratios, because of the reduced number of size fractions. Thus, judgement of the benefit of one fine aggregate against another needs to take account of the combined effects of changes in mean size and voids ratio, and the predominant level of cement content likely to be involved.

There has been a historic mistrust of fine sands in some parts of the UK. This may have been due to comparisons being made on concretes of constant sand content rather than after adjustments to ensure optimised cohesion.

On the other hand, in localities where fine sands are in common use, high quality concretes are made routinely, without problems, by appropriate design. The appropriate design is simply to use less of a fine sand. This accords with the theoretical model which results in reduced proportions of fine to total aggregate when fine sands are used.

This also accords with the literature. For example, Banfill and Carr (1987) demonstrated that concretes made with very fine sands can be designed properly without need for high water contents. Kantha Rao and Krishna-moothy (1993) identified that the use of sands of a restricted range of sizes reduced particle interference. Kronlof (1994) described the benefit of using very fine aggregate in superplasticised concrete to reduce water demand through improved particle packing.

Glanville *et al.* (1938) concluded that the inclusion of up to 40% crusher dust in fine aggregate had no direct effect on strength, its prime effect being on water demand and that substantial percentages could be included without detriment.

Combined theoretical effects of mean size and voids ratio of fine aggregate on the water demand of concrete

The combined theoretical effects of mean size and voids ratio of fine aggregate have been investigated for the situation when all individual size fractions have the same value of 0.85 for the voids ratio.

Table 7.2 Properties of BS 882 aggregates

BS 882 designat'n	C	M	F	F*
Mean size (mm)	0.9	0.63	0.45	0.24
Voids ratio	0.46	0.50	0.57	0.72

The four fine aggregate gradings selected for the comparison were three gradings central within the BS 882 designations C, M and F, plus a fourth grading at the fine limit of F and labelled F*. The mean sizes and voids ratios were estimated by successive theoretical blending of the finest size with the next size until all size fractions had been blended in the desired proportions. The results are summarised in Table 7.2.

The values for water demand and per cent fine/total aggregate are summarised in Figure 7.9, for the four fine aggregates, from estimates by computer simulation assuming a typical Portland cement and 20 mm graded coarse aggregate.

It will be observed that there is little difference in water demand for the M, F and F* fine aggregates for the leaner concretes, but the finer gradings show an advantage in medium and rich concretes. Fine aggregate C, despite its lower voids ratio has a substantially higher water demand over the entire

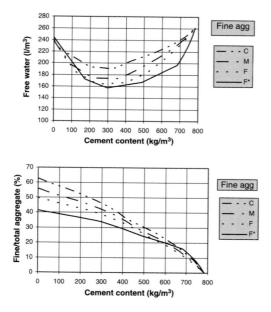

Figure 7.9 Theoretical combined influences of mean size and voids ratio on water demand and per cent fine/total aggregate for the range of gradings permitted in BS 882.

cement content range due to adverse effect of its relatively high mean size on particle interference.

The fine/total aggregate ratio has been decreased automatically for the finer fine aggregates, the effect being greater towards the leaner concretes. The direction and magnitude of the changes accord with normal concreting experience.

Different voids ratios and proportions of the individual size fractions may well produce different comparisons, particularly if the voids ratios of the size fractions also differ.

7.2.4 Influence of per cent fines on the water demand of concrete

The theoretical effects were examined for two sands of using lower and higher percentages of fine/total aggregate than the safe optimum values. The results are shown in Figure 7.10, from which it may be seen that, for both sands, an increase of 10 and 15 in the per cent fines would be required before the water demands were increased by more than 5 and 10 l/m^3 respectively. Thus, moderate increases in the per cent fines to increase cohesion should not yield serious effects on water demand and strength. On the other hand, decreasing the per cent fines by more than 5, say, below the safe optimum value would lead to reduced cohesion and a high rate of increase in water demand more particularly for the finer sand. This accords with the view of Kirkham (1965) who reported that the water demand of a fine sand was more sensitive to changes in percentage of fine aggregate than for a coarser sand.

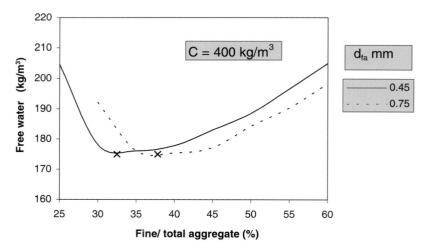

Figure 7.10 Example of the theoretical effects of sand mean size and per cent fines on the water demand of concrete. The points marked x are the safe optimum values of per cent fines.

Table 7.3 Example of effect of cohesion factor on per cent fines and water demand of a particular concrete

c 350	d_{fa} 0.6	u_{fa} 0.65
Cohesion factor	Per cent fines	Water (l/m^3)
0	34	174
1	37	174
2	41	175
3	44	177

7.2.5 Influence on concrete water demand of the selected value of cohesion factor

A typical example is provided in Table 7.3, illustrating the small effects on water demand of concrete resulting from changing cohesion factor, within the range 0 to 3. Reasons for such changes might include allowance for concrete pumping, variation in grading of the aggregates or achievement of a particular surface finish.

Use of a value between 0 and 1 would normally only be relevant at low workabilities, or when examination of the concrete indicates that the concrete is over-cohesive. This latter situation might apply if the fine or coarse aggregate is abrading or degrading in the mixer, leading to a finer sand or higher per cent fines than assumed in the design.

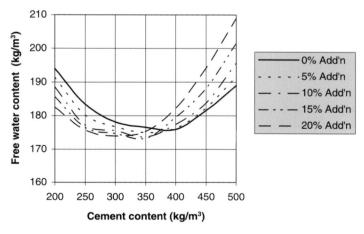

Figure 7.11 Influence of water demand of concrete, of inert fine material addition of 35 μm mean size.

7.2.6 *Influence of fillers or aggregates of intermediate size between cements and fine aggregates*

Figure 7.11 demonstrates from theory, the benefit to leaner concretes of the inclusion of a fine material intermediate in size between cement and aggregates. In this example, the fine material has a mean size of 35 µm, the size of fine dust from crushing of rock, or silt from a natural sand/gravel deposit with clay removed. The filler and the cement are assumed to have the same voids ratio of 0.895

It will be seen that there is substantial benefit to water demand for lean and medium cement content concretes but in this example, above a cement content of about 380 kg/m^3, there is sufficient fine material in the form of cement, so that the addition of more fine material is not beneficial and may be detrimental.[2]

7.3 Overview

Theoretical effects of changes in properties of materials on the proportioning of aggregates and the water demand of concrete are in close agreement with data and experience from research and practice.

8   Case studies

During the course of the research, opportunities arose to demonstrate the Theory of Particle Mixtures as an investigative tool. Selected extracts from reports of some of the topics investigated are included here as case studies. They are not necessarily either comprehensive or finalised.

8.1 A mineral addition for concrete

The following analysis is concerned only with assessment of properties of fresh pastes in the Vicat test using particular samples of cement and a very fine addition. The theoretical values of voids ratio and water demand in Table 8.1 and Figure 8.1 have been made by application of the Theory of Particle Mixtures.[1,2]

The closeness between the values in columns 6 and 7 in Table 8.1 suggests that

(a) The mean size of the addition particles has been under-estimated, and/or

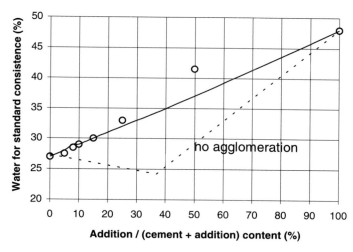

Figure 8.1 Comparison between theoretical and observed water contents of cement paste.

Table 8.1 Comparison between theoretical water contents for standard consistence and measured values

1	2	3	4	5	6	7
Addition content % by mass of cement plus add'n	RD of powder	Void ratio if size of add'n is 0.007 mm	Water for SC % if size of add'n is 0.007 mm	Void ratio if size of add'n is 0.015 mm	Water for SC % if size of add'n is 0.015 mm	Observed water for SC %
0	3.2	0.89	27	0.89	27	27
5	3.16	0.87	27	0.92	28	27.5
8	3.14	0.86	27	0.93	29	28.5
10	3.13	0.85	26	0.94	29	29
15	3.09	0.83	26	0.96	30	30
25	3.03	0.79	25	1.01	32	33
50	2.87	0.89	29	1.12	37	41.5
100	2.6	1.32	48	1.32	48	48

(b) There has been additional agglomeration or some equivalent effect in the mixtures.

In effect, the presence of even a small proportion of addition has resulted in the cement as well as the addition suffering from agglomeration or some equivalent effect with no benefit from void filling by the smaller sized material.

From Figure 8.1, it may be seen that there is considerable benefit to be obtained for water demand, by reduction in particle interference and/or agglomeration, if the actual size or the effective size of the addition in paste can be reduced to 7 μm or less. This may require higher energy mixing or the inclusion of a plasticiser.

8.2 Investigation of fine aggregate performance

On occasion materials perform differently from expectations in concrete. The most common explanations are

• The materials properties are different from those expected or measured, due to sampling effects or time-dependent changes.
• The material includes adsorptive components, e.g. certain clays such as montmorillonite
• The process of mixing of concrete modifies the grading and voids ratio of the material due to attrition or fragmentation.

In a recent case, anomalous results were obtained during a project for the Advanced Concrete Technology Diploma of the Institute of Concrete

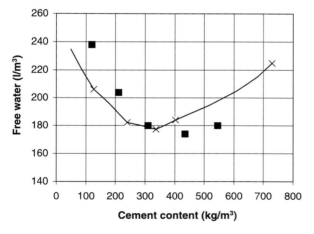

Figure 8.2 Comparison between predicted and experimental water demands of concrete using unadjusted data for the fine aggregate.

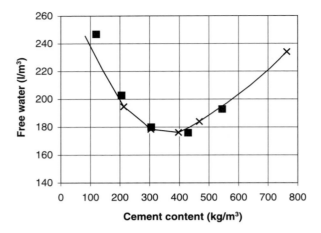

Figure 8.3 As for Figure 8.2 but using adjusted data for the fine aggregate.

Technology, when the candidate was comparing the results of simulation based on materials properties with results of trial mixes in the laboratory.

Sufficient data were available to eliminate cement and coarse aggregate from the enquiry and to focus attention on the fine aggregate.

The mean size and voids ratio obtained from the fine aggregate tests were 0.855 mm and 0.590 respectively, but use of these values led to the misfit shown in Figure 8.2 when computer simulation was attempted.

Adjustments to the values for mean size and voids ratio of the sand suggested that more appropriate values would be 0.675 mm and 0.825 respectively which would imply that the sand was reduced significantly in mean size and that the void ratio had increased as a consequence. The effect of using the adjusted values in computer simulation is shown in Figure 8.3.

A possible cause may be that weakly cemented sandstone particles present in the fine aggregate were breaking down in the mixing process.

8.3 Design of concrete incorporating heavyweight aggregates

Heavyweight aggregates may be required for special purposes, e.g. radio-active shielding. The densities of the materials are much higher than normal and the shapes and gradings of the materials may be different from those in normal use. As a result, *ad-hoc* adjustments to normal designs can be subject to appreciable uncertainty unless laboratory trials are made. Computer simulation could reduce design time and the risk of error.

An example of a computer simulated design by Minelco Ltd, using the Theory of Particle Mixtures, for counterweights made with heavy aggregates of 27.5 mm maximum size is shown in Table 8.2

Table 8.2 A design for heavyweight concrete

Materials	Batch weights (kg)
Cement	210
Water	125
Fine agg (SSD)	1190
Coarse agg (SSD)	4150
Total	5675

8.4 An assessment of the concept of an ideal grading curve for aggregates or for concrete

Powers(1968) and Popovics (1979) have reviewed critically the historic development of 'ideal' grading curves, including

Fuller parabolic grading
Bolomey grading
Faury grading

Powers concluded, 'The hypothesis that there is an ideal size gradation for concrete aggregates, or for all the solid material in concrete has now become almost if not entirely abandoned.'

Popovics commented that 'It is unrealistic to expect that there is any single grading that can optimise all, or most, concrete properties simultaneously', that '... it is little wonder that none of the ideal gradings ... has been proved optimum from the standpoint of concrete technology' and further that 'The optimum grading, as well as the optimum quantity of sand in concrete aggregate, depend on, among other things, the grading and type of coarse aggregate used. The same is true in a reverse sense for the optimum grading of coarse aggregate.'

Earlier, Abrams (1924) referred to '... the absurdity of our [i.e. USA – 1924] present practice in specifying definite gradings for aggregates ...'

More recently, Lees (1970a) concluded that there can be no such thing as an ideal grading curve. Day (1995) in discussing the ideal grading curves of Fuller and Thompson and of Bolomey concluded that these approaches have two basic weaknesses

(i) It is rarely possible to replicate the gradings in the field
(ii) The ideal grading for one use could not simultaneously be ideal for all uses.

The prevailing views above are confirmed by the Theory of Particle Mixtures, in that for minimum water demand it is necessary to vary the proportioning of the fine and coarse aggregates to take account of other properties of the aggregates as well as their gradings and also the cement properties and cement content of the concrete. Thus, despite the continuing use in some countries of ideal curves for gradings of aggregates, the concept of a single ideal overall aggregate grading is unlikely to be profitable and will not be pursued further.

As a corollary, it would not be logical for the complex interaction of void filling and particle interference to lead to a single ideal particle size distribution for concrete. Such distributions take into account only the size range, and not the shape and texture of particles, which also affect the void filling capabilities of each size fraction. The only instance when such a concept might be valid could be when all sizes of the complete distribution for concrete are available for use and when each size has the same or sensibly similar value of voids ratio to that applying to the materials used by the supporter of such a concept.

On the other hand, Plum (1950) argues that the concrete grading, i.e. the combined grading of all the components, should be taken into account in concrete design. Popovics (1979) recommended characterising the overall distribution for the complete concrete by a method which depends only on a knowledge of the cement content and the largest size of aggregate. Again, this proposal is suggested as unlikely to be valid for all situations, because it does not take account of the voids ratios of the individual size fractions of the materials, but at least it might be a substantial improvement over methods for the aggregates alone.

The formula proposed by Popovics for P the cumulative percentage passing each sieve size is

$$P = g + (100 - g) \times k^b \tag{8.1}$$

where $g = 100/(A/C + 1)$, A is aggregate and C is cement, $b = 0.5$ and $k =$ sieve size/maximum size.

To assess this formula, the method developed from the Theory of Particle Mixtures for multiple components was adopted as follows

- Assessing the voids ratio of a combination of 11 component sizes of materials
- Optimising the proportions of each to produce a minimum voids ratio

utilising a Microsoft Excel spreadsheet and the Microsoft Solver technique.

Table 8.3 shows the 11 materials (coded 3 to 13) in the left-hand section, having a particle size distribution for the concrete conforming to Popovics criteria when 15% of cement is to be included by mass of the solid components, the void ratio of each size fraction is 0.80 and the relative densities are 3.2 and 2.6 for the cement and aggregates respectively.

In the right-hand section, the results are shown of the successive combination of each material, starting with material 3, using the formulae

Table 8.3 Assessment of the void ratio of a combination of 11 materials having a continuous overall distribution conforming to Popovics (1979) criteria

Material	Min. size input	Max size input	Void ratio U input	Rel dens RD input	Per cent by mass input	Mean size D	Per cent by vol	Rel densy of comb'n	Void ratio of comb'n	Mean size of comb'n
1	0.0025	0.005	0.8	3.2	0	0.004	0		0.800	
2	0.005	0.01	0.8	3.2	0	0.007	0		0.800	
3	0.01	0.02	0.8	3.2	15	0.014	13		0.800	
4	0.02	0.04	0.8	2.6	1	0.028	1		0.774	
5	0.04	0.08	0.8	2.6	2	0.057	2		0.708	
6	0.08	0.15	0.8	2.6	2	0.110	2		0.644	
7	0.15	0.3	0.8	2.6	3	0.212	3		0.561	
8	0.3	0.6	0.8	2.6	4	0.424	4		0.478	
9	0.6	1.2	0.8	2.6	7	0.849	7		0.382	
10	1.2	2.5	0.8	2.6	9	1.732	9		0.307	
11	2.5	5	0.8	2.6	13	3.536	13		0.242	
12	5	10	0.8	2.6	18	7.071	19		0.193	
13	10	20	0.8	2.6	26	14.142	27		0.159	
14	20	40	0.8	2.6	0	28.284	0	2.68	0.159	1.683
				Sum	100	Sum	100	Output	Output	Output

Table 8.4 Modification of Table 8.3 as a result of optimising the distribution to minimise the voids ratio of the combination

Material	Min. size input	Max size input	Void ratio U input	Rel dens RD input	Per cent by mass input	Mean size D	Per cent by vol	Rel densy of comb'n	Void ratio of comb'n	Mean size of comb'n
1	0.0025	0.005	0.8	3.2	0	0.004	0		0.800	
2	0.005	0.01	0.8	3.2	0	0.007	0		0.800	
3	0.01	0.02	0.8	3.2	15	0.014	13		0.800	
4	0.02	0.04	0.8	2.6	0	0.028	0		0.800	
5	0.04	0.08	0.8	2.6	3	0.057	3		0.705	
6	0.08	0.15	0.8	2.6	3	0.110	3		0.616	
7	0.15	0.3	0.8	2.6	3	0.212	3		0.547	
8	0.3	0.6	0.8	2.6	5	0.424	5		0.448	
9	0.6	1.2	0.8	2.6	8	0.849	8		0.358	
10	1.2	2.5	0.8	2.6	10	1.732	11		0.286	
11	2.5	5	0.8	2.6	14	3.536	14		0.230	
12	5	10	0.8	2.6	17	7.071	18		0.188	
13	10	20	0.8	2.6	23	14.142	23		0.156	
14	20	40	0.8	2.6	0	28.284	0			
				Sum	100	Sum	100	2.68	0.156	1.511
					Sum		Sum	Output	Output	Output

of the Theory of Particle Mixtures in section 3.1. At the foot of the right-hand side, shown boxed, are the resultant properties of the total combination. Thus, the combination of 11 components, each having the same voids ratio of 0.80, resulted in a composite voids ratio of 0.159.

The exercise was repeated but utilising Microsoft Solver to modify the distribution to minimise the voids ratio. The results are shown in Table 8.4.

It will be seen from Tables 8.3 and 8.4 and Figure 8.4 that the effect of optimization was to reduce the voids ratio marginally from 0.159 to 0.156 by relatively minor changes to the distribution.

Thus, for this particular example, with all sizes available and assuming that the voids ratios are the same for all sizes, the Popovics method and the Theory of Particle Mixtures for optimized concrete yielded similar results for the particle size distribution and the voids ratio.

Further work is necessary to more fully compare results of the two methods.

8.4.1 Continuous gradings v gap gradings for aggregates for concrete

The available materials may not lend themselves to the production of a continuous distribution and there is a long history of the satisfactory use of gap-graded materials. Indeed there are strong supporters of the use of gap-graded materials because of perceived benefits.

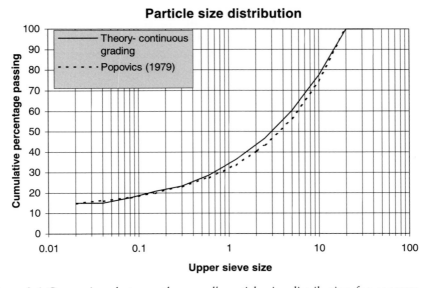

Figure 8.4 Comparison between the overall particle size distribution for concrete predicted by the Theory of Particle Mixtures, with that developed by Popovics for a continuous distribution.

For, example, Stewart (1962) considered that the fine aggregate should be sufficiently small so that the majority of particles would pass through the openings between coarse particles. This was necessary to ensure spaces were filled and that additional voids were not created by wedging apart of coarse particles. Stewart concluded that the provision of a gap in the aggregate grading to be most desirable.

Ball (1998) utilised grading parameters as a starting point for predicting optimum particle packing but warned that this does not take account of shape and may also lead to error when a significant gap occurs in the overall grading.

Lees (1970b) concluded that there was an advantage in having a number of gaps in the overall grading of a particulate material under the constraints that each successive lower size should be

(i)   Not so small that filtration is encouraged
(ii) Not so large as to cause the coarser particles to be unduly separated.

Ehrenburg (1980), supported by Li and Ramakrishnan (1983), demonstrated advantages in omitting the finer sizes of coarse aggregate, thus providing a substantial gap in the grading between the coarse and fine aggregate. Loedolff (1986) accepted the Fuller curve as a sound overall grading but identified that with materials resulting in gapped gradings it is necessary to adjust the final grading about the Fuller curve to provide a balanced result. de Larrard and Buil (1987) suggested that there are benefits to be gained from gaps in the over-all grading of concrete between the coarse aggregate, fine aggregate and cement. Kessler (1994) investigated the theoretical benefits of gap gradings and developed 3-dimensional mathematical models for combining successively smaller sizes. Although the model was based on single-sized spheres Kessler recognised that in reality other shapes occurred in nature and that exact single sizes were impractical.

On the other hand, Popovics (1979) concluded that selection of gap grading or continuous grading is made in most cases on an economical basis. Ehrenburg (1981) applied Weymouth's concepts of particle interference to the design of both continuously graded and gap graded concretes, implying that both can make good concrete. Day (1995) recognised that gap-graded concretes have a greater tendency to segregate than continuously graded concretes.

To examine gap gradings using the Theory of Particle Mixtures, 'double-size' gaps were created between the cement and sand and between the sand and coarse aggregate. The gap-graded particle size distribution was then optimized for minimum voids ratio for the same conditions applied to the continuous distribution.

It will be seen from Table 8.5 and Figure 8.5 that, in this particular case, the introduction of substantial gaps in the distribution did not lead to a benefit for voids ratio. Indeed the voids ratio is marginally greater than that

Table 8.5 Results of optimising the distribution in Table 8.4 for minimum voids ratio and providing gaps in the distribution between the cement and sand and between the sand and coarse aggregate

Material	Min. size input	Max size input	Void ratio U input	Rel dens RD input	Per cent by mass input	Mean size D	Per cent by vol	Rel densy of comb'n	Void ratio of comb'n	Mean size of comb'n
1	0.0025	0.005	0.8	3.2	0	0.004	0		0.800	
2	0.005	0.01	0.8	3.2	0	0.007	0		0.800	
3	0.01	0.02	0.8	3.2	15	0.014	13		0.800	
4	0.02	0.04	0.8	2.6	0	0.028	0		0.800	
5	0.04	0.08	0.8	2.6	0	0.057	0		0.800	
6	0.08	0.15	0.8	2.6	6	0.110	6		0.591	
7	0.15	0.3	0.8	2.6	2	0.212	2		0.538	
8	0.3	0.6	0.8	2.6	5	0.424	5		0.449	
9	0.6	1.2	0.8	2.6	7	0.849	8		0.360	
10	1.2	2.5	0.8	2.6	0	1.732	0		0.360	
11	2.5	5	0.8	2.6	0	3.536	0		0.360	
12	5	10	0.8	2.6	31	7.071	32		0.193	
13	10	20	0.8	2.6	34	14.142	35		0.159	
14	20	40	0.8	2.6	0	28.284	0	2.68	0.159	2.176
				Sum	100	Sum	100	Output	Output	Output

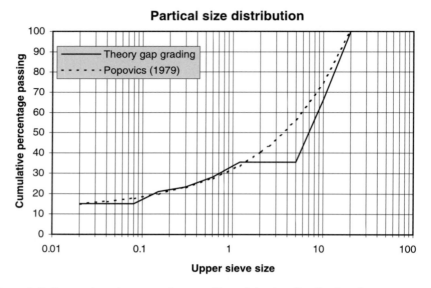

Figure 8.5 Comparison between the overall particle size distribution for concrete predicted by the Theory of Particle Mixtures for a gap graded concrete, with a continuous distribution as developed by Popovics (1979).

obtained for the optimized continuous distribution in Table 8.4. However, it would be reasonable to conclude that in this one example, both continuous and gap-graded distributions yielded sensibly similar voids ratios such that either could be used for concrete, the final choice being affected by such practical factors as economy and availability.

This pilot investigation does not imply that all gap gradings and continuous gradings will necessarily produce similar results. If the sizes to be omitted are those having the highest voids ratios, due to poor shape or texture, then a benefit is more likely. Similarly, if the fine aggregate or coarse aggregate has a voids ratio which may benefit from the omission of one or more sizes then a gap grading may prove very beneficial. Each case would need to be considered on its merits.

Some perceived benefits, favouring either continuous gradings or gap gradings, may stem from completely valid but particular experiences which may then have led to an unjustifiable generalisation favouring one or the other type. Another reason could be that comparisons are made without ensuring that the proportioning has been adjusted, to take account of the changes needed in the overall grading when changing from one type to the other. For the optimized gap-grading, the proportion of fine to total aggregate was 24% whereas for the continuous gradings the proportion was about 50%.

Of course, none of this takes account of performance in particular practical situations. Much of the perceived benefit attached to gap grading relates to ease of compaction under vibration, whereas those favouring continuous gradings identify less sensitivity to variations and less proneness to segregation under vibration. Such other factors as materials availability and cost may be the most significant factors affecting the final choice of gap or continuous grading.

The scale of the observed differences in overall grading in Figure 8.5 confirm the experience of practitioners who are well aware that there is a very wide range of acceptable distributions, both continuous and gap-graded, which will result in economic concrete, provided the correct proportioning is achieved in each case.

8.4.2 Design of concrete to match the materials or design of materials to match the concrete. Are both concepts valid? What is the role of computer simulation?

The discussion in section 8.4.1 leads to consideration of the concepts of

- Designing concretes to make the best use of the available materials
- Designing materials to enable the best use of the concrete

and the role that models, such as the Theory of Particle Mixtures, can play in achieving the best solutions for either purpose.

Producers of materials for concrete will opt for general purpose solutions to cover the majority of their market and may additionally provide special purpose solutions to cover niche markets.

Producers of concrete, likewise, will stock general purpose materials to cover the majority of their market and, when necessary, may stock special materials to cover niche markets. Additionally, most producers will optimise the selection and proportioning of the materials for minimum cost to meet client specifications.

As more materials are made available and concrete producers attempt to respond, the opportunities increase to fine-tune their selection and proportioning but the costs of the fine-tuning increase because of design time involved.

Additionally, as more materials are introduced it is necessary to have more accurate information available for control and adjustment purposes to satisfy quality assurance requirements.

Thus, the role of models, theories and related techniques is to enable designs to be made fast, accurately and economically and to provide the relationships to assist process control.

This does not imply that testing is not needed. As suggested by Dewar (1998, 1992) it is preferable to have

Table 8.6 Properties of individual size fractions and of the resultant combination for the initial cement

1	2	3	4	5	6	7	8	Combination		
								9	10	11
Material	Min. size (mm)	Max size (mm)	Void ratio U	Rel dens RD	Per cent by mass	Mean size D (mm)	Per cent by vol	Mean size (mm)	Void ratio	Rel density
1	0.0025	0.005	1.8	3.2	15	0.004	15		1.80	
2	0.005	0.01	1.5	3.2	25	0.007	25		1.35	
3	0.01	0.02	1.3	3.2	25	0.014	25		1.11	
4	0.02	0.04	1.2	3.2	35	0.028	35		0.91	
5	0.04	0.08	1.1	3.2	0	0.057	0		0.91	
6	0.08	0.15	1.0	3.2	0	0.110	0	0.012	0.91	3.20

- Forward control of materials to predict changes in relationships
- Immediate control of the product during production
- Retrospective control of the product to confirm that the process is operating correctly

with built-in feedback loops as necessary.

For the materials producers, materials cannot be designed without consideration of how they will perform in combination with other material in their clients' products. Models, theories and computerised simulation enable fast answers to 'what-if' type scenarios, enabling the most likely best solution to be predicted before confirming in the laboratory, in pilot production or in full-scale production. Thus, as the options multiply, computerised simulation provides an essential tool in the kit of the developer.

8.5 Further examples of optimisation of materials and concretes

In this section further examples are provided following the theme of the previous section concerning design of materials and concretes.

In the various examples, the voids ratios adopted for the individual size fractions are selected to be typical values and to yield realistic properties of the combinations. However, they should not be considered to be more than guidance values and would need to be determined by Vicat test or loose bulk density test, as appropriate for each size fraction of the particular material. It is to be expected that, as assumed, the voids ratio of the finer materials will increase with decreasing size due to adverse influences of shape and interparticle forces.

In all cases, the percentage of fine to total aggregate is adjusted for a cohesion factor of 1.

8.5.1 Optimisation of particle size distribution of the cementitious material for minimum water demand of concrete at different cement contents

This particular example concentrates on optimisation of a cementitious material for minimum water demand of concrete and does not consider any effects on the strength potential of the material caused by changes to the particle size distribution.

The initial material is considered in Table 8.6 in terms of convenient size fractions in columns 1–3 for which the voids ratios and relative densities, together with the proportions by mass, are listed in columns 4 to 6. The mean size of each size fraction in column 7 has been calculated on a logarithmic basis and the volumetric proportions are shown in column 8.

Material 1 is combined mathematically with material 2, using the author's Theory of Particle Mixtures (Dewar 1983, 1986, 1997). The resultant voids

Table 8.7 Properties of the optimised cement for minimum water demand of concrete at a cement content of 400 kg/m³

| 1 | 2 | 3 | 4 | 5 | 6 | 7 | 8 | Combination | | |
								9	10	11
Material	Min. size (mm)	Max size (mm)	Void ratio U	Rel dens RD	Per cent by mass	Mean size D (mm)	Per cent by vol	Mean size (mm)	Void ratio	Rel density
1	0.0025	0.005	1.8	3.2	14	0.004	14		1.80	
2	0.005	0.01	1.5	3.2	9	0.007	9		1.47	
3	0.01	0.02	1.3	3.2	21	0.014	21		1.09	
4	0.02	0.04	1.2	3.2	40	0.028	40		0.84	
5	0.04	0.08	1.1	3.2	16	0.057	16		0.76	
6	0.08	0.15	1.0	3.2	0	0.110	0	0.018	0.76	3.20

ratio, 1.35, is recorded in column 10. Similarly the mean size and relative density of this combination (omitted for clarity) are calculated. This combination is then combined with Material 3 and so on until all size fractions have been combined. It will be seen that the voids ratio progressively decreases as each size fraction is added. The final properties of the combination are shown at the foot of the table.

To optimise the proportions in column 6, the resulting properties for the cementitious material are fed into the computerised concrete design programme (see example in Table 8.12) and the water demand is estimated for selected slump, cement content and a particular set of aggregates. The values in column 6 are then allowed to vary systematically within the computer program until a minimum concrete water demand is established. The results are summarised in Table 8.7, from which it may be seen that, on average, optimisation suggests that a coarser cement is to be preferred at this cement content, with a reduced voids ratio in consequence, compared with that of the finer cement selected initially.[3]

The results of optimising the cement at different cement contents are summarised in Table 8.8 from which it may be seen that the initial cement has properties closer to those for the optimised cements over the lower part of the cement content range.

At high cement contents a coarser cement is to be preferred with regard to water demand. Of course other constraints, e.g. bleeding, strength, and economy may determine whether this is feasible.

Table 8.8 Properties of cements optimised for concrete water demand at different cement contents

Cement kg/m^3	Cement properties		
	Mean size	Void ratio	Rel dens'y
	Initial cement		
All	0.012	0.91	3.20
	Optimised cements		
150	0.008	1.19	3.20
200	0.017	0.79	3.20
250	0.017	0.77	3.20
300	0.013	0.88	3.20
350	0.017	0.78	3.20
400	0.018	0.76	3.20
450	0.023	0.72	3.20
500	0.020	0.74	3.20
550	0.023	0.72	3.20

The overall range for the particle size distributions for the optimised cements is shown in Figure 8.6 and the benefit for water demand in Figure 8.7.

The optimum proportion of the finest material (5–2.5 μm) did not vary significantly from the maximum constraint value of 15%, implying that a higher percentage could be beneficial and that material finer than 2.5 μm might also be beneficial. The average optimised percentage of the coarsest material (40–80 μm) was 18% and ranged from 0% at the lowest cement content to 35% for the highest cement content.

Examination of Figure 8.7 indicates that, for this example, the initial cement was effectively optimised for particle size distribution with respect to water demand over the lower part of the working range of cement content but was less well optimised for higher cement contents. There is a potential reduction in water demand of up to 15 l/m³ to be achieved by adopting a

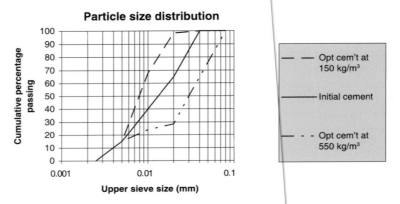

Figure 8.6 The range of particle size distributions of cements optimised for minimum water demand of concrete at different cement contents.

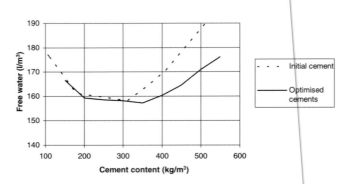

Figure 8.7 Influence of optimisation of particle size distribution of cement on the water demand of concrete.

coarser psd for the cementitious material at high cement contents. How such optimisation might be achieved is considered later.

These findings are consistent with the often disappointing results of using more finely ground cements, which have resulted in higher water demands of high cement content concretes. The usual reasons are that finer size fractions have greater voidage and also that the benefit of reduced particle interference has been offset by the narrow psd which has resulted in a higher voids ratio and thus increased water demand. Conversely, low cement content concretes benefit more from the reduced particle interference by the use of finer cements and may, as a consequence enable slightly reduced water demands.

8.5.2 *Optimisation of particle size distribution of the fine aggregate for minimum water demand of concrete at different cement contents*

The optimisation of fine aggregate proceeds as for cement, commencing as in Table 8.9 for an initial fine aggregate which lies centrally in the range of permitted gradings in BS 882: 1992. The cement used for this exercise is that adopted initially in the previous exercise.

To optimise the proportions in column 6, the resulting properties for the fine aggregate are fed into the computerised concrete design programme and the water demand is estimated. The values in column 6 are then allowed to vary systematically within the computer program until a minimum concrete water demand is established at the selected cement content. The results are shown in Table 8.10, from which it may be seen by comparison with Table 8.9 that a significantly finer grading is to be preferred, despite an increased voids ratio, compared with that for the fine aggregate selected initially.[4]

The results of optimising the fine aggregate at different cement contents are summarised in Table 8.11 from which it may be seen that over the complete range of cement contents from $150-550$ kg/m^3, a significantly finer grading was to be preferred compared with that of the initial fine aggregate, despite the increased voids ratio. There is possibly some indication of a trend favouring higher mean size at lower cement contents.

The gradings for the optimised fine aggregates are summarised in Figure 8.8 in comparison with that for the initial fine aggregate. It will be seen that the range of the optimised gradings is relatively narrow and does not overlap the original grading.

It is probable that relaxation of the constraint of a maximum of 4% passing 0.08 mm (80 μm) would further modify the optimisation. This aspect is considered further later.

Figure 8.9 demonstrates that, in this particular example, optimising the fine aggregate grading enables a substantial reduction of water demand, in the order of $15-35$ l/m^3, over the full working range of cement content.

Table 8.9 Properties of individual size fractions and of the resultant combination for the initial fine aggregate

1	2	3	4	5	6	7	8	9	10	11
								Combination		
Material	Min. size (mm)	Max size (mm)	Void ratio U	Rel dens RD	Per cent by mass	Mean size D (mm)	Per cent by vol	Mean size (mm)	Void ratio	Rel density
5	0.04	0.08	1.1	2.6	2	0.057	2		1.10	
6	0.08	0.15	1.0	2.6	8	0.110	8		0.95	
7	0.15	0.3	0.9	2.6	15	0.212	15		0.79	
8	0.3	0.6	0.7	2.6	20	0.424	20		0.63	
9	0.6	1.2	0.7	2.6	25	0.849	25		0.54	
10	1.2	2.5	0.7	2.6	20	1.732	20		0.47	
11	2.5	5	0.7	2.6	10	3.536	10	0.64	0.44	2.60

Table 8.10 The average optimised fine aggregate for a cement content of 400 kg/m³

1	2	3	4	5	6	7	8	9	10	11
								Combination		
Material	Min. size (mm)	Max size (mm)	Void ratio U	Rel dens RD	Per cent by mass	Mean size D (mm)	Per cent by vol	Mean size (mm)	Void ratio	Rel density
5	0.04	0.08	1.1	2.6	4	0.057	4		1.10	
6	0.08	0.15	1.0	2.6	39	0.110	39		0.98	
7	0.15	0.3	0.9	2.6	13	0.212	13		0.89	
8	0.3	0.6	0.7	2.6	15	0.424	15		0.76	
9	0.6	1.2	0.7	2.6	18	0.849	18		0.64	
10	1.2	2.5	0.7	2.6	11	1.732	11		0.59	
11	2.5	5	0.7	2.6	0	3.536	0	0.28	0.59	2.60

8.5.3 Options for optimisation of cements and fine aggregates

It is now appropriate to consider the options in order to benefit from the exercise so far.

Of course, the cement and aggregate industries have long experience in the successful optimisation of their materials with regard to minimising production costs, improving properties and developing market benefits for their products both with regard to general purpose aggregates and also products for the larger niche markets. However, cements are restricted by the

Table 8.11 Properties of sands optimised for use at different cement contents

Cement kg/m³	Fine aggregates		
	Mean size	Void ratio	Rel dens'y
	Initial sand		
All	0.64	0.44	2.60
	Optimised sands		
150	0.33	0.54	2.60
200	0.33	0.54	2.60
250	0.24	0.63	2.60
300	0.22	0.67	2.60
350	0.22	0.66	2.60
400	0.28	0.59	2.60
450	0.22	0.67	2.60
500	0.21	0.67	2.60
550	0.17	0.75	2.60

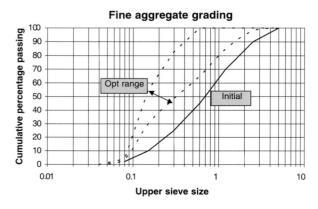

Figure 8.8 Optimised fine aggregate gradings for minimum water demand of concrete.

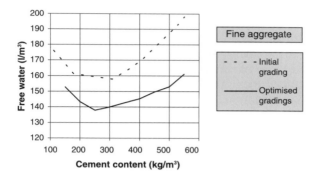

Figure 8.9 Influence of optimisation of fine aggregate grading on the water demand of concrete.

need to meet other requirements, e.g. strength, and aggregates are significantly restricted by the constituents, shape and grading provided in nature or as a result of processing. In addition, specifications may place restrictions upon the grading of aggregates, in particular on the proportion of very fine material, especially if it likely to contain clay rather than silt or dust.

However, there always remains considerable scope for significant general improvements to voids ratio and modifications to psd or grading, as well as modifications for the specialised needs of smaller niche markets.

For cements in particular, it would be beneficial to gain improvements in both voids ratio and psd at the higher cement contents.

Methods by which improvements may be obtained include any of the following taken separately or in combination

- Improvements in the voids ratios and/or gradings of the materials
- Blending with materials having more favourable shape, surface characteristics or enabling a more favourable grading.
- Addition of very fine material to extend the size distribution at the lower end, e.g. mineral filler or other powder.
- Addition of material intermediate in size between the cement and fine aggregate.

With regard to each of these, it will be observed that in the examples already described, constraints were deliberately applied, so that advantage was not taken of some of these possibilities. In subsequent sections, some of these options and possible benefits will be considered in more detail.

It will be appreciated that the opportunities for making improvements are not restricted to a single party. For example, modification of grading at the interface between cement and fine aggregate could result from efforts by cement producer, addition supplier, aggregate supplier or concrete producer.

8.5.4 Inclusion of mineral fillers in the range of sizes between cement and sand

Cement producers normally try to minimise the coarse material content in cement because of the benefit to strength. Aggregate producers are constrained by BS 882 to a maximum of 4% silt or fine dust in a gravel sand and 16% in a crushed rock sand.

However, the analysis so far suggests that a significant benefit could be obtained by appropriate inclusion of material in the size range between cement and fine aggregate. Of course, if it was intended to introduce a filler of say 80–40 µm, say, it would be necessary to take account of any cement above 40 µm and any fine aggregate below 80 µm.

Further analyses have been made to evaluate the optimum percentage of such a filler in the concretes investigated earlier using the initial cement and a fine sand. The results are summarised in Figure 8.10 from which it may be seen that reductions of about 5 l/m³ in free water content are suggested to be attainable by including 23% filler (80–40 µm) with a sand which contained only the permitted maximum 4% passing 80 µm.

Naturally, if the cement also contained a significant proportion of material coarser than 40 µm, this would reduce the benefit of including fillers.

8.5.5 Optimisation of particle size distribution of the coarse aggregate for minimum water demand of concrete at different cement contents

There is normally a significant benefit to be gained for water demand by adopting the largest possible maximum size of coarse aggregate. However, maximum size is usually restricted by specified requirements determined by construction conditions. In the examples in this paper, 20 mm maximum size has been adopted.

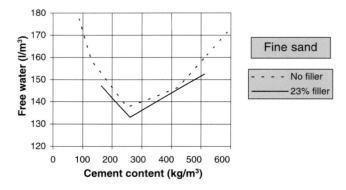

Figure 8.10 Influence on concrete water demand of the inclusion of 23% mineral filler with a particular fine aggregate.

With regard to grading of the coarse aggregate, when two sizes, 20–10 mm and 10–5 mm are available both having the same voids ratio, computer analyses as in Figure 8.11 have indicated that with medium sands there is a benefit to water demand in using less than 50% of the 10 mm size. For fine sands, 30–50% of 10 mm was optimal for the complete range of cement content.

If, as is often the case, the 10 mm voids ratio is higher than that for the 20 mm material, the optimum proportion would be reduced.

8.5.6 Introduction of material intermediate in size between the fine and coarse aggregate

By implication from the analyses it would rarely be advantageous to introduce intermediate sizes between the fine and coarse aggregates unless the voids ratio of the introduced material was very low or the initial fine aggregate was very fine.

Thus, introduction of a medium or coarse sand to supplement a very fine sand might only very occasionally be beneficial.

8.5.7 Inclusion of cementitious additions or mineral fillers in the range of cement sizes or in finer sizes

It is well documented that the inclusion of fine or very fine materials can often reduce the water demands of concretes and provide other benefits, e.g. economy; strength; thermal properties; durability. Some materials, due to advantageous shape and surface texture may assist in reducing the voids ratio of the cementitious material generally. Others may assist in extending the range of sizes with a similar benefit.

As recorded earlier, the optimum percentage of the finest cement size used, 5–2.5 μm equalled the constraint value of max 15%, implying benefit of

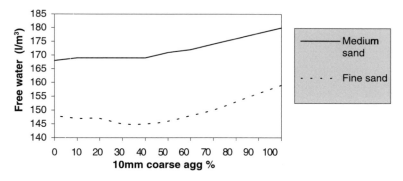

Figure 8.11 Example of influence on water demand of the percentage of 10 mm material in 20 mm coarse aggregate in concrete of 400 kg/m³ cement content.

Table 8.12 Input data for computerised simulation of concrete using 7 of 9
particulate materials and a superplasticiser

Summary	Title	Input	Example		
Materials	Input	Mean size mm D Input	Void ratio U Input	Void ratio U Adjusted	Rel Densy RD Input
Cem 1	Addn 1				
Mass %	8	0.0015	2.100		2.2
vol %	10.3				
Cem 2	Addn 2				
Mass %	25	0.010	0.700		2.3
vol %	30.7				
Cem 3	PC				
Mass %	67	0.0120	0.910		3.2
vol %	59.1				
Cement & Additions		0.0092	0.743	0.632	2.82
Fine ag 1	Filler				
mass (%)	23	0.06	1.100		2.6
vol %	23.0				
Fine ag 2	Fine sand				
mass (%)	77	0.24	0.630		2.6
vol %	77.0				
Fine ag 3	Med Sand				
mass (%)	0	0.64	0.440		2.6
vol %	0.0				
Fine agg & minl addn		0.174	0.623	0.529	2.60
CAgg1	10 mm				
% mass	30	7.07	0.80		2.60
% vol	30.0				
CAgg2	20 mmSS				
% mass	70	14.14	0.80		2.60
% vol	70.0				
CAgg3	40 mmSS				
% mass	0	26.5	0.80		2.60
% vol	0.0				
Coarse aggregate		11.49	0.726	0.617	2.60

Conditions					Input
Cohesion factor		0.5 to 2	Normally 1		1
			Air cohesion factor		1
Air content (%)		Total	Normally 0.5, 1 or 5.5		1
		Entrapped	Assume 0.5–1.5		1
		Entrained			0
Mortar		Input Y for mortar mixtures			N
Slump mm		0 to 250	Normally 75 or 50		50
Admixture	Code				Sprplasr
	Rate	Litres/100 kg cement			0.75
	Factor	0.5 to 1	Use 1 if no water redn or no admixture used		0.85
Cement strength		28d			55
strength factor		28 day	factor		1
		7 day	factor		1.15
aggreg factor					0.95
Cemaddn fr	Addn 1	Str efficiency factor		0 to 3	2
	Addn 2	Str efficiency factor		0 to 3	0.35
	PC	Str efficiency factor		0 to 3	1

using at least 15% below 5 μm and possibly a percentage of material below 2.5 μm.

Obviously, if the particular material had a very high voids ratio, its benefit might be limited despite the benefit to particle interference of the smaller size. In addition, the small size might be accompanied by a tendency for agglomeration such that the size of agglomerated particles was much larger than the size of individual particles and they might resist separation during mixing. Furthermore the presence of this powder when mixed with cement particles might transmit the agglomeration effect to the cement as well. Of course there may be other benefits to be taken into account, e.g. greater reactivity and strength-producing qualities, which offset any detrimental qualities.

Additional analyses were made to indicate whether experience would be confirmed by theory. This also provides the opportunity to display the input and output data involved in the author's system of concrete simulation for both fresh and hardened concrete.

Table 8.12 shows the materials input data for 9 particulate materials of which 7 have been selected for use in the example. The top box shows a very fine cementitious material having a high voids ratio, a medium material with a low voids ratio and the initial cement used previously. The proportions have been selected at common levels. The resultant properties of the combination

Table 8.13 Computerized simulation of concrete using 7 of 9 particulate materials and a superplasticiser to meet a particular specification

Specification	Input	cem + adn (kg/m³)	Selected concrete to meet specification			Output
Min cement kg/m³	300	300	Cem + cem addn		kg	342
Max free w/c	0.35	294	Addn		kg	27
Max aggregate/cement	6	333	Addn 2		kg	85
Char strength N/mm²	65		PC		kg	229
Plant sd N/mm²	4		Free water incl admix		kg	103
Margin factor	2.00		Admixture	Sprplasr	litres	2.6
Margin N/mm²	8		Filler	ssd	kg	146
Design mean strength	73	342	Fine sand	ssd	kg	490
N/mm²	1		Med sand	ssd	kg	0
Total air %			10 mm	ssd	kg	407
Slump	50		20 mmSS	ssd	kg	949
Per cent of material permitted by the specification to be called cement			40 mmSS	ssd	kg	0
			Plastic density		kg/m³	2436
	For min	For max	For	Free w/(c + adn)			0.30
	cement	a/c	w/c	Aggregate/(cement + adn)			5.82
Addn 1	100	100	100	% fines incl min adn			32
Addn 2	100	100	100	Total air		%	1
PC	100	100	100	Expected mean str		N/mm²	73
				Materials cost		£/m³	

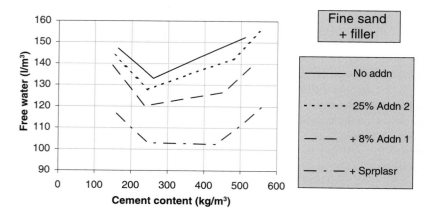

Figure 8.12 Comparison between the water demands for concretes containing progressively more components.

are shown at the foot of the box. The adjusted voids ratio allows for the effect of the superplasticiser.

The second box shows a mineral filler and fine sand selected at the percentages found to produce minimum concrete water demand. The medium sand is not used for this exercise. The third box shows the 10 mm and 20 mm coarse aggregates selected at percentages in the range found to produce minimum concrete water demand. The 40 mm aggregate is not used for this exercise.

The final box provides for a number of important additional items of data about the materials and the concrete, including the air content, required slump, admixture and strength factors. In this case, a superplasticiser has been included.

An example specification and output data from the simulation are provided in Table 8.13. In this example, the strength is the key requirement determining cement content. The concrete quantities and properties from the simulation process are listed on the right.

Figure 8.12 demonstrates the benefits to water demand and strength of successive inclusion of more materials, including a superplasticiser.

9 A user-friendly computerised system for general application

This section illustrates sequentially, examples of computer screen displays of Mixsim98 which has been developed and marketed for use in a Microsoft Windows environment by Questjay Limited, on the basis of the author's modelling system for fresh and hardened concrete.

9.1 Materials database

The **Materials database** (Figure 9.1) summarises the details of the particulate **Materials** and the liquid **Admixtures**. The Materials in the database are available for use in any number of concrete mixture simulations.

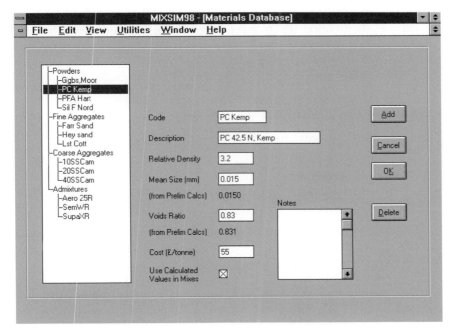

Figure 9.1 Screen display – materials database.

Four types of Materials are identified,

- Powders – cements and cementitious additions
- Fine aggregates – sands, crushed fine material, and mineral fillers
- Coarse aggregates – gravels and crushed rock
- Liquid admixtures

For the particulate materials, the **Code**, **Description**, **Relative density** and **Cost**, if known, of each new material, are entered in the right-hand section. If the **Voids ratio** and **Mean size** are also known already, or are being assumed, the values may also be entered. Alternatively, the necessary properties may be calculated from the basic test data.

9.2 Preliminary calculations – general

The **Preliminary Calculations Screens** enable the test data for each material to be input and then converted to the properties needed by MIXSIM98.

9.2.1 Preliminary calculations – powders – cements and additions

The screen (Figure 9.2) shows the input data and resulting output properties for a typical Portland cement.

The **Mean size** of cement or other powders may be determined from the particle size distribution (PSD) by entering the percentages of the material passing each tabulated sieve size. If other sieve sizes than those tabulated have been used, it will be necessary to interpolate the percentages at the tabulated sizes. Alternatively a special set of sieve sizes can be created.

9.2.2 Preliminary calculations – aggregates

The screen shows an entry for a fine aggregate (Figure 9.3).[1]

9.3 Materials screen

Materials names and properties may be entered by either

1. Selection from the **Materials** database, using 'drag and drop' technique, or
2. Direct entry to the **Materials** screen.

The first method is appropriate when intending to prepare batch data and designs for materials whose properties have been pre-determined. The second method is appropriate when exploring the effects of changing properties of materials or when transferring data from other sources.[2]

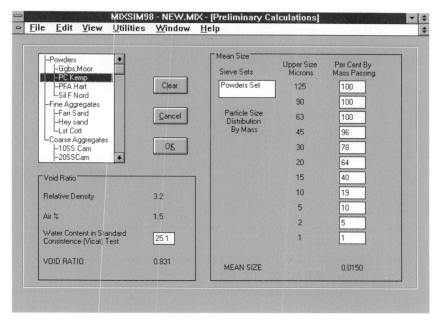

Figure 9.2 Screen display – preliminary calculations – cement properties.

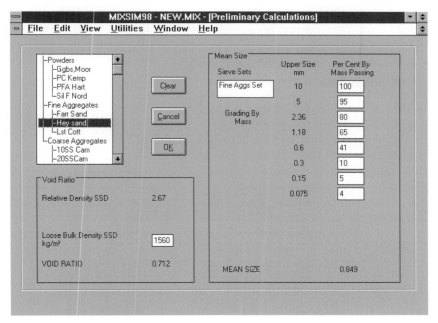

Figure 9.3 Screen display – preliminary calculations – fine aggregate properties.

MIXSIM98 automatically calculates the properties of the combination of each Type of material and displays the results at the foot of each section (Figure 9.4). Also shown in the materials Code column in the three sections are the **Adjusted voids ratios** allowing for any reduction, as a result of the selection of a **Plasticiser** and entry of a corresponding **Admixture factor** via the **Conditions screen**.

9.4 Mix conditions

There are a number of decisions to be made concerning **Mix conditions** (Figure 9.5) including

- Cohesion factor
- Total air content
- Entrapped air content
- Slump (mm) at the time of delivery
- Admixture code(s) dosage rates and factors
- Cement strength – 28 d is the EN196 mortar prism strength at 28 days.
- Aggregate strength factor
- Strength factor – 28 d is normally 1
- Strength factors – 7 d and 56 days

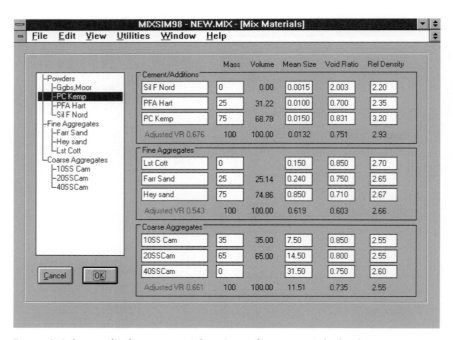

Figure 9.4 Screen display – materials – input from materials database.

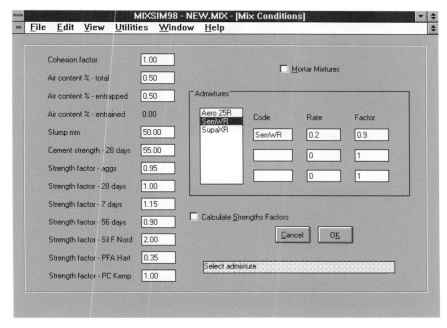

Figure 9.5 Screen display – conditions including admixtures.

- Strength factors – 28 d for the powders
- Mortar mixtures.

9.5 Concrete specification

The left-hand side of the **Concrete specification screen** allows input of a number of common specification requirements affecting the cement content and cost of the concrete (Figure 9.6). These are

- Specification code
- Minimum cement plus additions, i.e. powder content
- Maximum aggregate/cement plus additions ratio
- Maximum free water/cement plus additions ratio
- Minimum 28 day characteristic strength
- Concrete strength – standard deviation usually 2–5 N/mm^2
- Strength margin factor usually 2–2.5

At the bottom left of the screen is shown a table allowing input of the percentage of each cementitious material permitted by the specification to

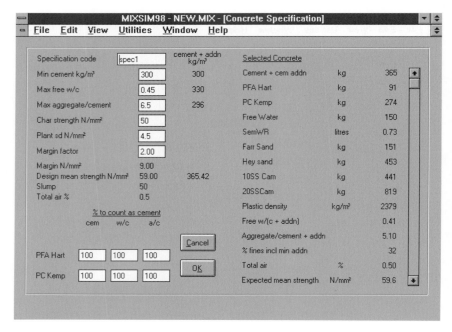

Figure 9.6 Screen display – concrete specification and selected concrete (see also Figure 9.7).

count towards cement content, aggregate/cement and water/cement for durability.

The right-hand side of the **Concrete specification screen** displays the quantities of materials and properties of the **Selected concrete** designed to satisfy the specified requirements from the left-hand side of the screen and the materials and conditions shown on the **Materials** and **Conditions screens**.

The primary specification aspect determining the final choice of concrete can be identified in the central section of the screen which shows the (cement + additions) contents needed to meet the various specified requirements. The highest (cement + additions) content is adopted in the right-hand section and is a key factor determining the quantities, properties and cost of the selected concrete.

Comparison between the left and right-hand sides of the screen, scrolling as necessary, will show that all the specified requirements are expected to be met by the selected concrete.

If a scroll bar is to be seen at the right-hand side of the screen (Figure 9.7) some aspects of the selected concrete can only be viewed by scrolling the screen down.

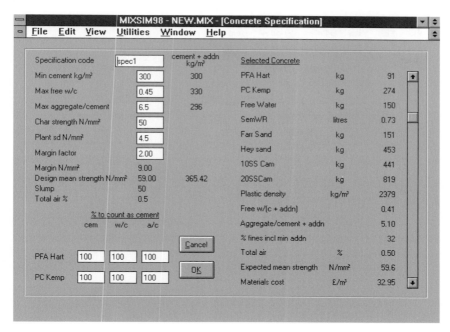

Figure 9.7 Screen display – concrete specification and selected concrete, demonstrating the use of the scroll bar to view additional information (see also Figure 9.6).

9.6 Concrete trials data

Up to eight sets of concrete trial data may be input (Figure 9.8) including

- Cement plus addition content, i.e. Powder content kg/m^3
- Free water content kg/m^3 or litres/m^3
- % fine/total aggregate
- Plastic density kg/m^3
- Slump mm
- Total air content %
- Strength N/mm^2 at 28 days and also at 7 and 56 days.

The ratios of free water/(cement + addition) content and of aggregate/ (cement + addition) content are calculated automatically.

9.7 View

The **View** screens enable stages in the calculations to be viewed for

- Detailed assessments of the effects of changes, or

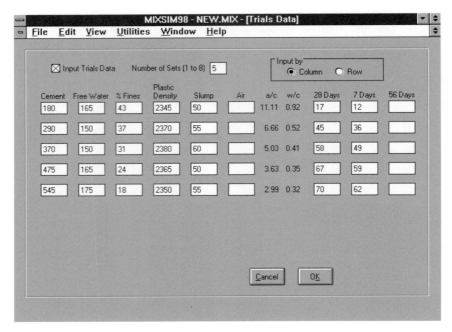

Figure 9.8 Screen display – concrete trial data.

- Educational purposes, or
- General interest

in the intricacies of the many detailed parts of the simulation processes, particularly when the full range of 9 particulate materials and 3 admixtures is involved.

The **View** screens are divided into 3 sections, and each section into a number of separate aspects. The 3 sections are

- Simulation stages
- Results tables
- Graphs

The most important in everyday use will be the later screens of **Results tables**, which cover the tables used for calculating batch data, and the **Graphs** which compare simulated concrete data and any concrete trial data.

Each of the sections of Results tables (Figures 9.9–9.11) has a similar vertical arrangement in that the proportions and properties of 17 concretes are shown consisting of 6 mixtures, a–f, together with 11 interpolated mixtures.

MIXSIM98 - NEW.MIX - [Results Tables]

File   Edit   View   Utilities   Window   Help

Volumes of voids and solids
Detailed volume proportions of voids and solids
Mass of cement + additions + water + admixture

Pt	Cem + Add	PFA Hart	PC Kemp	Water	SemWR	Farr Sand	Hey sand	10SS Cam	20SSCam	Density	Fines %
A	0	0	0	241	0.00	261	782	324	602	2209	52.97
	24	6	18	227	0.05	255	766	335	623	2232	51.60
	49	12	37	214	0.10	250	751	347	645	2256	50.21
	73	18	55	199	0.15	245	735	360	668	2280	48.81
B	98	24	73	184	0.20	240	720	373	692	2306	47.40
	128	32	96	177	0.26	232	695	381	707	2320	46.00
	159	40	119	170	0.32	223	670	389	722	2333	44.58
	189	47	142	162	0.38	215	646	397	738	2348	43.14
C	218	55	164	155	0.44	207	622	406	754	2362	41.68
	281	70	211	151	0.56	185	554	420	780	2371	38.07
D	340	85	255	148	0.68	163	488	435	807	2381	34.40
	390	97	292	151	0.78	140	419	447	830	2378	30.44
E	434	108	325	155	0.87	118	355	459	853	2374	26.50
	529	132	397	171	1.06	83	250	463	859	2356	20.12
	611	153	458	187	1.22	52	156	466	866	2337	13.51
	678	170	509	202	1.36	24	73	469	871	2318	6.76
F	733	183	550	216	1.47	0	0	472	877	2298	0.00

Figure 9.9 Screen display – results tables – materials quantities per m^3 of concrete.

MIXSIM98 - NEW.MIX - [Results Tables]

File   Edit   View   Utilities   Window   Help

Mass of cement + additions + water + admixture
Aggregates in SSD condition
Density and proportions

Pt	Density	Fines %	w/c	a/c	7 Day Strength	28 Day Strength	56 Day Strength	PFA Hart	PC Kemp	SemWR	Farr Sand	H
A	2209	52.97	9.35	81.40	0	0	0	0.00	0.00	0.00	1.82	
	2232	51.60	9.35	81.40	0	0	0	0.27	1.00	0.05	1.79	
	2256	50.21	4.38	40.92	0	0	0	0.55	2.01	0.10	1.75	
	2280	48.81	2.72	27.44	0	0	0	0.82	3.02	0.15	1.72	
B	2306	47.40	1.89	20.72	1	1	2	1.10	4.03	0.20	1.68	
	2320	46.00	1.38	15.69	3	5	7	1.44	5.29	0.26	1.62	
	2333	44.58	1.07	12.63	7	11	14	1.79	6.55	0.32	1.56	
	2348	43.14	0.86	10.58	13	18	23	2.12	7.78	0.38	1.51	
C	2362	41.68	0.71	9.11	21	27	33	2.46	9.01	0.44	1.45	
	2371	38.07	0.54	6.89	35	43	49	3.17	11.61	0.56	1.29	
D	2381	34.40	0.44	5.58	47	56	62	3.82	14.01	0.68	1.14	
	2378	30.44	0.39	4.71	53	62	69	4.39	16.09	0.78	0.98	
E	2374	26.50	0.36	4.12	57	65	71	4.88	17.89	0.87	0.83	
	2356	20.12	0.32	3.13	60	68	74	5.96	21.84	1.06	0.58	
	2337	13.51	0.31	2.52	62	70	76	6.87	25.18	1.22	0.36	
	2318	6.76	0.30	2.12	63	71	77	7.63	27.97	1.36	0.17	
F	2298	0.00	0.29	1.84	63	71	77	8.24	30.23	1.47	0.00	

Figure 9.10 Screen display – results tables – properties of concrete.

Pt	PFA Hart	PC Kemp	SemWR	Farr Sand	Hey sand	10SS Carr	20SSCarr	Total Cost
A	0.00	0.00	0.00	1.82	4.69	2.59	4.21	13.32
	0.27	1.00	0.05	1.79	4.60	2.68	4.36	14.76
	0.55	2.01	0.10	1.75	4.50	2.78	4.52	16.21
	0.82	3.02	0.15	1.72	4.41	2.88	4.68	17.67
B	1.10	4.03	0.20	1.68	4.32	2.98	4.84	19.14
	1.44	5.29	0.26	1.62	4.17	3.05	4.95	20.78
	1.79	6.55	0.32	1.56	4.02	3.11	5.06	22.40
	2.12	7.78	0.38	1.51	3.88	3.18	5.17	24.01
C	2.46	9.01	0.44	1.45	3.73	3.25	5.28	25.61
	3.17	11.61	0.56	1.29	3.32	3.36	5.46	28.77
D	3.82	14.01	0.68	1.14	2.93	3.48	5.65	31.70
	4.39	16.09	0.78	0.98	2.52	3.58	5.81	34.14
E	4.88	17.89	0.87	0.83	2.13	3.67	5.97	36.24
	5.96	21.84	1.06	0.58	1.50	3.70	6.02	40.65
	6.87	25.18	1.22	0.36	0.94	3.73	6.06	44.36
	7.63	27.97	1.36	0.17	0.44	3.75	6.10	47.42
F	8.24	30.23	1.47	0.00	0.00	3.78	6.14	49.86

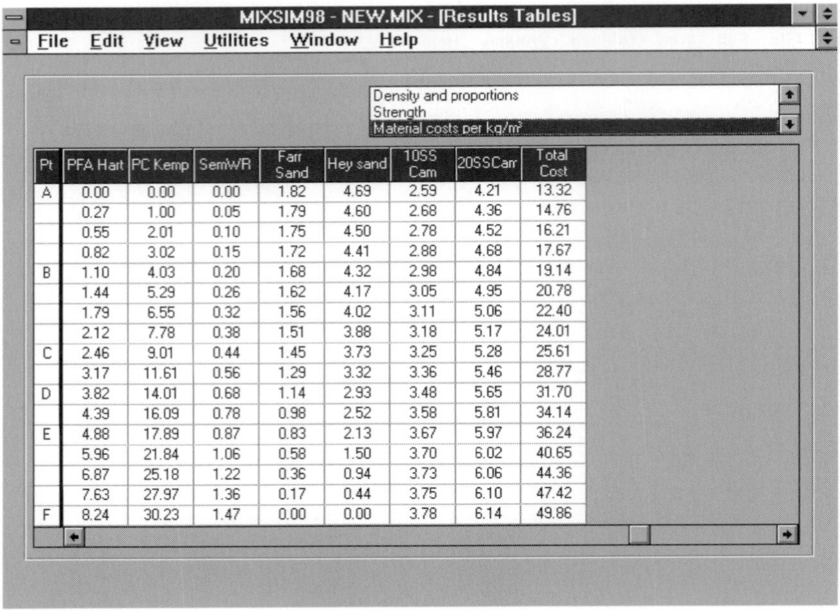

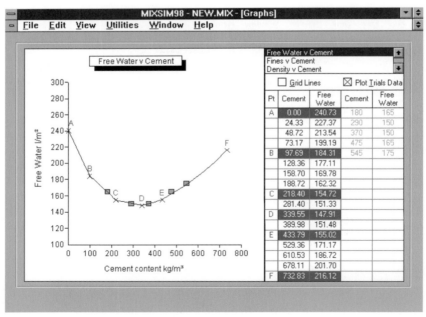

Figure 9.11 Screen display – results tables – materials costs of concrete.

Pt	Cement	Free Water	Cement	Free Water
A	0.00	240.73	180	165
	24.33	227.37	290	150
	48.72	213.54	370	150
	73.17	199.19	475	165
B	97.69	184.31	545	175
	128.36	177.11		
	158.70	169.78		
	188.72	162.32		
C	218.40	154.72		
	281.40	151.33		
D	339.55	147.91		
	389.98	151.48		
E	433.79	155.02		
	529.36	171.17		
	610.53	186.72		
	678.11	201.70		
F	732.83	216.12		

Figure 9.12 Screen display – water demand of concrete.

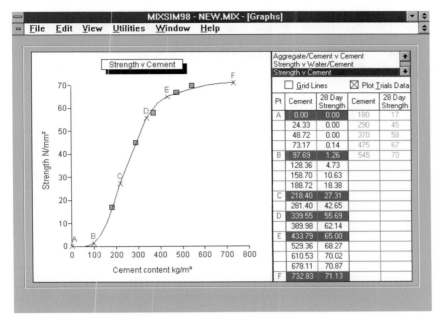

MIXSIM98 - NEW.MIX - [Graphs]
File Edit View Utilities Window Help

Strength v Cement

Aggregate/Cement v Cement
Strength v Water/Cement
Strength v Cement

☐ Grid Lines   ☒ Plot Trials Data

Pt	Cement	28 Day Strength	Cement	28 Day Strength
A	0.00	0.00	180	17
	24.33	0.00	290	45
	48.72	0.00	370	58
	73.17	0.14	475	67
B	97.69	1.26	545	70
	128.36	4.73		
	158.70	10.63		
	188.72	18.38		
C	218.40	27.31		
	281.40	42.65		
D	339.55	55.69		
	389.98	62.14		
E	433.79	65.00		
	529.36	68.27		
	610.53	70.02		
	678.11	70.87		
F	732.83	71.13		

Figure 9.13 Graphs – 28 day strength v cement content of concrete.

9.8 Graphs

Graphs of the following are provided (Figures 9.12 and 9.13)

- Water demand v cement content
- Per cent fines v cement content
- Plastic density v cement content
- Free water/cement v cement content
- Aggregate/cement v cement content
- 28 d Strength v free water/cement
- 28 d Strength v cement content
- 28 d Strength v 7 d strength
- Materials cost v cement content

There are options to show trial data and grid lines. Graphs may be viewed full screen size by double clicking the graph area.

9.9 Mixes database

One of the most important and useful facilities of MIXSIM98 is the option to compare the results of using different materials or different proportions for the same specification (Figure 9.14).

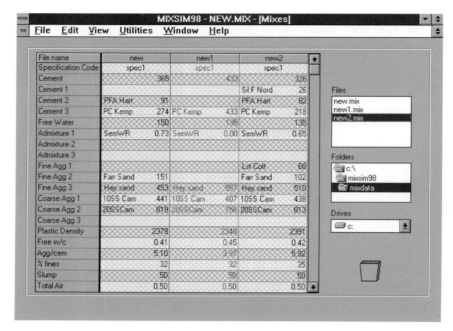

Figure 9.14 Screen display – mixes database for comparing concretes.

The **View** menu may be used to select the **Mixes database**. By selecting mixes from the right-hand box and using 'drag and drop' technique, the mixes may be compared in the blank columns on the left, scrolling as necessary if more than three mixes are to be compared.

If a scroll bar is to be seen at the right-hand side of the screen some aspects of the selected concrete can only be viewed by scrolling the screen down.

9.10 Utilities

Utilities contains screens enabling

- Sets of Sieve sizes to be added or modified
- Program settings to be selected
- Strength factors to be calculated
- Batching data to be obtained.

9.10.1 *Batching data*

This screen enables Batching data for each material to be viewed and printed out for full-size batches of concrete allowing for the moisture contents of the aggregates (Figure 9.15).

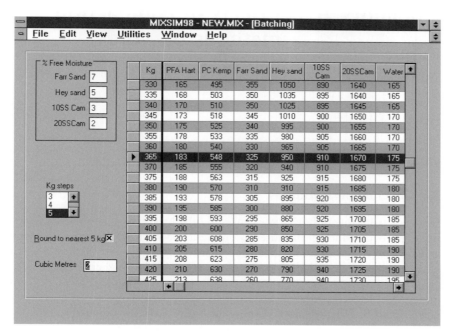

Figure 9.15 Screen display – batchbook.

Appendix A   Effects of compaction methods on voids ratios

A1 General considerations

Evidence from the author's experience and the literature generally, suggested that the behaviour of mixtures of materials would vary under different compaction conditions. Obviously, application of higher energy should result in a lower resultant voids ratio but, if the effect of that higher energy was to cause or increase segregation, the reduction in voids ratio might not materialise or be less than expected.

Materials having a relatively high voids ratio in the uncompacted state due to high surface friction or angular shape might compact to a much greater extent under high energy compaction compared with smooth round materials. Indeed, it was shown by the work of Moncrieff (1953) that rounder and smoother material showed relatively poor benefit from higher energy compaction compared with angular rough material.

It will be apparent from discussion later, that some methods of compaction used in research work may have resulted in segregation that has not been identified. The purpose of this appendix is to suggest where such segregation may have occurred and to quantify the effects.

It is important that the method of compaction should reasonably relate to practical conditions and this might require different methods for different products.

With regard to conventional ready mixed concrete, it is the author's experience that the use of rodding or vibration of the aggregates led to unrealistically higher densities than were obtained in concrete even under normal vibration. Exceptions might include very low workability concretes, subject to intense vibration and pressure, and also grouted concretes involving pre-placed vibrated aggregates.

A2 Effect of compaction energy level on voids ratios

The author's experimental data on low energy compaction are presented in section 6.1.

Table A1 Data for combinations of single sized and multi-sized aggregates under high energy compaction (intense vibration) obtained by Loedolff (1986)

Materials code LL	J5	J6	J7	J8	J9	J10	J11	J12	J13	J14	J15	J16
D mm	5.4	9.25	12.3	12.3	12.3	11	15.5	15.5	15.5	14.5	15.5	10
d mm	0.11	0.74	0.11	0.74	5.40	0.47	8.50	12.3	1.9	5.40	5.4	0.475
Adjustment factor F	1	1	0.85	1	1.15	1	0.9	1	1	1	1	0.95
Size ratio $r = Fd/D$	0.020	0.080	0.008	0.060	0.505	0.043	0.494	0.79	0.12	0.37	0.35	0.045

Prop'n of fine material

Voids ratio   U

n	J5	J6	J7	J8	J9	J10	J11	J12	J13	J14	J15	J16
0	0.710	0.700	0.750	0.740	0.740	0.630	0.810	0.820	0.820	0.740	0.820	0.58
0.10	0.580	0.590										
0.21	0.450	0.510	0.500	0.500	0.660	0.430	0.710	0.770		0.650	0.630	0.4
0.31	0.370	0.410	0.340	0.420	0.640	0.370	0.650	0.740	0.450	0.590	0.600	0.33
0.41	0.400	0.340	0.350	0.310	0.600	0.280	0.650	0.740	0.380	0.590	0.580	0.27
0.51	0.460	0.330	0.400	0.300	0.620	0.240	0.600	0.740	0.310	0.580	0.570	0.26
0.61	0.530	0.330	0.450	0.270	0.620	0.250	0.580	0.740	0.260	0.590	0.580	0.25
0.71	0.630	0.320	0.550	0.300	0.620	0.260	0.590	0.750	0.270	0.600	0.600	0.25
0.81			0.600	0.370	0.640	0.280	0.620	0.750	0.240	0.620	0.660	0.26
1	0.760	0.370	0.760		0.700	0.350	0.620	0.740		0.710	0.700	0.35

Table A2 Data of de Larrard et al. (1987, 1994) for combinations coded LB of single sized aggregates under high energy compaction (vibration), and LS R21 for aggregates compacted under vibration and pressure plate

Materials code	LB R54ad	LB R54af	LB R54be	LB R54bex	LB R54cd	LB R54cf	LS R21
D mm	2.25	8.7	4.41	4.41	2.25	8.7	8
d mm	0.313	0.313	0.682	0.682	1.22	1.22	0.50
Adjustment factor F	0.85	1	1	1	0.95	1.2	1
Size ratio r = Fd/D	0.118	0.036	0.155	0.155	0.515	0.168	0.063
Prop'n of fine material n	*Voids ratio* U						
0	0.560	0.610	0.600	0.560	0.550	0.610	0.590
0.10	0.440	0.500	0.490	0.490	0.540	0.540	0.470
0.20	0.370	0.430	0.430	0.410	0.515	0.470	0.370
0.30	0.330	0.370	0.380	0.350	0.500	0.390	0.320
0.40	0.330	0.330	0.360	0.330	0.490	0.370	0.330
0.50	0.370	0.330	0.370	0.370	0.490	0.370	0.350
0.60	0.410	0.360	0.390	0.390	0.500	0.390	0.390
0.70	0.450	0.400	0.440	0.430	0.515	0.410	0.470
0.80	0.490	0.470	0.480	0.470	0.540	0.450	0.540
0.9	0.560	0.515	0.530	0.515	0.560	0.490	0.610
1	0.61	0.57	0.59	0.56	0.575	0.56	0.67

Data on high energy compaction have been analysed from the work of Loedolff (1986) and of de Larrard *et al.* (1987–1994) for single-sized or multi-sized materials in combination.

The data of Loedolff are summarised in Table A1 and of De Larrard *et al.* in Table A2. The data are plotted as void ratio diagrams in Figures A1 and A2 in comparison with theoretical diagrams based on the Theory of Particle

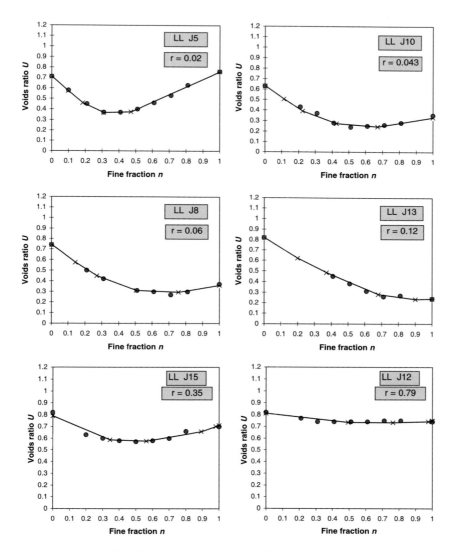

Figure A1 Examples of comparisons between theoretical voids ratio diagrams and experimental data for high energy (HE) compaction using data obtained by Loedolff (1986).

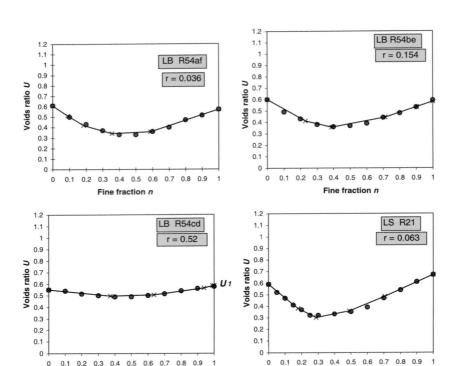

Figure A2 Examples of comparisons between theoretical voids ratio diagrams and experimental data for high energy (HE) compaction using data quoted by de Larrard (1987, 1994).

Mixtures and data for **low energy** compaction, but **with modified constants** to allow for the higher energy conditions, as discussed in section A3. Adjustments have been made to some of the values for size ratio, r, to improve the fit; the adjustment factors are also shown in the data tables.

A3 Relationships between z, the void band-width factor, and r, the size ratio

Estimates of z, the voids band-width factor, were made by the method given in section 3.1.3 for the author's series of aggregate tests at low energy levels for compaction and for those of Loedolff (1986) and Sedran *et al.* (1994) at high energy levels.

A3.1 Low energy levels

The results for z from the author's work are shown plotted against r in Figure A3. As z is influenced by U_0, relationships are shown for the two values of U_0 representing the most typical high and low values.

It is possibly logical that when the size ratio r approaches zero, z should also be zero because particle interference is eliminated. However, the data support this only partially. The most probable reason is that for the very small sizes of particles, the effective particle size and voids ratio are inflated by interparticle forces. Alternatively, it is possible that the relatively small extent of segregation observed even under low energy compaction at low r values has had an influence.

Curves have been fitted empirically to the data. No theory has been involved other than to ensure that the values of z when $r = 1$ are consistent with the concept of the change points lying on the line joining U_0 and U_1 in the void ratio diagram. This requires, for $n = 1$, that

$$z(n = 1) = (1 + U_0)^{0.33} - 1 \qquad (A1)$$

This implies that z increases with U_0, and is consistent with detailed examination of the data.

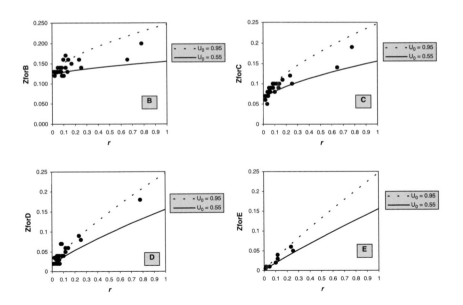

Figure A3 Influence of size ratio r and voids ratio U_0 of the coarser component on the Z values for aggregate mixtures at the change points B–E in voids ratio diagrams for low energy (LE) compaction in the JDD series.

The selected formula is

$$z = b_1 + [(1 + U_0)^{0.33} - 1 - b_1]r^{b^2} \tag{A2}$$

This formula results in a series of shallow domed curves for z between the values at $r = 0$ and 1, the position of the curve being influenced by U_0.

A3.2 High energy levels

The void ratio diagrams for data of Loedolff (1986), de Larrard and Sedran (1994) and Sedran *et al.* (1994) and de Larrard and Buil (1987) have also been analysed for values of z. The relationships are illustrated in Figure A4 and the constants in the formula are shown in Table A3 in comparison with those from the author's series for the lower energy compaction.

The constants b_1 and b_2 in the formula for the four change points $B-E$ are summarised in Table A3.

The data in Figure A4, obtained under high energy compaction, generally show trends for increasing z with increasing r, but also show a substantial discontinuity in the relationship at low values of r. The points of discontinuity, indicated by the arrows, are at an r value of approximately

$$0.14 \ U_0^{0.33} \quad \text{i.e. between } r = 0.11 \text{ and } 0.14 \text{ for } U_0 \text{ in the range } 0.55 \text{ to } 0.95.$$

The one exception is the set of results obtained for z from the work of de Larrard and Sedran (1994). In this case pressure was applied to the top surface of the mixture. It is probable that this pressure has restricted the movement of coarse particles and enabled an r value as low as 0.05 to be used under high energy compaction, without significant segregation occurring.

These aspects have not been referred to in the published papers of Loedolff or of de Larrard and co-workers. However, they are consistent with a computerised simulation by Barker (1994) for vibrated mixtures, illustrated in

Table A3 Constants in the formula for z, assuming no significant segregation

Constants	Compaction energy level			
	Low JDD Series		High LL, LB, LS series	
Points	Intercept b_1	Power b_2	Intercept b_1	Power b_2
B	0.120	0.600	0.080	0.600
C	0.060	0.650	0.025	0.650
D	0.015	0.800	0.000	0.800
E	0.000	0.900	0.000	0.900

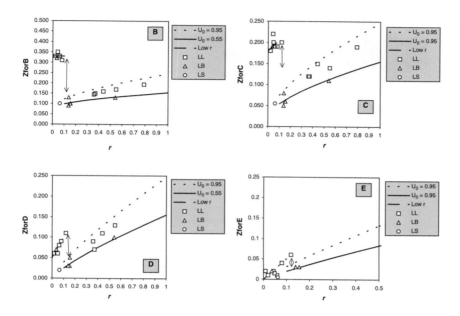

Figure A4 Influence of size ratio r and void ratio U_0 of the coarser component on the Z values for aggregate mixtures at the change points B–E in voids ratio diagrams for high energy (HE) level compaction in Series LL, LB and LS.

Figure A5 which shows partial segregation even with $r = 0.24$ and almost total segregation at $r = 0.1$.

Leutwyler (1993) reports that vibration can encourage convection and the rise of larger particles towards the surface where they are trapped, leading to a segregated mixture. In confirmation, the author some 30 years ago observed steel balls rising to the surface, and remaining there, in a vibrated mixture of dry cement powder and 12 mm steel balls.

A3.3 Comparison between effects of low and high energy levels

Higher energy compaction should lead to a reduction in voids ratio provided significant segregation is minimised. Thus, the component voids ratios, U_0 and U_1 should reduce under higher energy, as also should the voids ratios of mixtures. It may also be expected that z values for the void band-width factor at the change points should also reduce because higher energy compaction should enable mixtures of different sizes to benefit to a greater extent than either of the component materials.

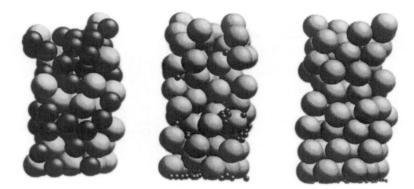

Figure A5 Computer generated simulation by Barker (1994) of stable sequentially
constructed packing of binary mixtures of single sizes of spheres under
the influence of vibration. The size ratios of the mixtures, from left to
right, are 0.80, 0.24 and 0.10.

Comparisons of effects of low and high energy compaction on the value of z
are shown in Figure A6. The arrows in each diagram indicate the
approximate position of the observed discontinuity when segregation occurs
for low r values under high energy compaction. For this example, it has been
assumed arbitrarily that the values of U_0 are 0.75 and 0.70 respectively for the
low and high energy conditions.

Thus, the benefits of higher energy compaction can be lost if segregation
occurs resulting in a large increase in the void band-width factor z. This is
much less likely to occur at values of r exceeding about 0.25 or perhaps 0.1,
because the fine particles acting singly, or in concert, will block the movement
through the gaps between the large particles.

For aggregates, in the condition when segregation occurs at low r values the
formula for estimating z is assumed to be linear, with the following equation
and the constants given in Table A4.

$$z = b_3 + b_4 \times r \tag{A3}$$

If the relationships found for aggregates are to have relevance for mortar and
concrete, it is important that they have not been influenced by adopting test
methods likely to produce results that are not mirrored in practice. Thus, if it
is intended that the methods of concrete manufacture should not lead to
significant segregation, as would be the normal case, then the methods of
testing the materials and the formulae should reflect this intention.

For most practical concreting situations, when care is taken to avoid
significant segregation, the relationships and constants suggested by the
experimentation for low r values under high energy compaction should
normally not be of relevance.

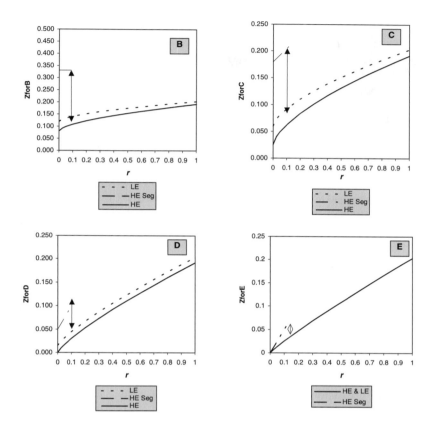

Figure A6 Influence of the level of compaction energy on the Z value for each of the four change points B–E, indicating the effect of low (LE) and high (HE) energy and the effect of segregation at low r values. U_0 has been taken arbitrarily as 0.75 and 0.70 for the LE and HE conditions respectively.

Table A4 Constants in the formula for estimating z when segregation occurs at low r values

Point	Intercept b_3	Slope b_4
B	0.33	0
C	0.18	0.25
D	0.05	0.5
E	0	0.5

A4 Overview of effects of energy level of compaction

The energy level adopted for compaction of particulate materials affects the voids ratio of mixtures. Provided segregation is avoided, the Theory of Particle Mixtures can be used but with modified constants in the formulae, appropriate to the method.

Under vibration, segregation may occur at low size ratios because the smaller size material moves down through the gaps in the coarse particles and the coarse particles may rise through convection. As a result, there will be a discontinuity in the relationship between void bandwidth factor and size ratio, and the composite voids ratio will be increased at low size ratios compared with the non-segregated condition.

If pressure is applied to the coarse particles during vibration, e.g. by the use of a plate, the size ratio at which segregation occurs may be reduced. The pressure from super-incumbent concrete may explain why segregation is less apparent in practice, compared with that which is deduced to have occurred with vibrated dry aggregate mixtures in the laboratory.

It is thus important to ensure that the test methods used reflect the situation applying in practice if the results are to be meaningful.

Unless the practical situation suggests that high energy vibration is absolutely essential for testing aggregates for bulk density and voids ratio, it is recommended that loose poured packing is adopted.

Appendix B   Adjustment of input data

When output data from theory and practice are compared, individual values can be expected to differ due to normal variations of input and output data.

Over the course of numerous assessments, the theory and the constants have been refined so that possibility of errors from these sources has been minimised, but of course cannot be wholly eliminated. On the other hand, assuming the validity of theory and the values of mathematical constants, any observed systematic differences suggest the possibility of differences in properties of materials between those tested and those actually used.

Effects of sampling and test method are considered to be the most likely causes of systematic discrepancies in the reported work, in situations when

- Materials input data are from certificates of test of bulk average samples from production while the laboratory trials are made with particular samples of materials assumed but not checked to be equally representative of the bulk
- A test method involved is relatively insensitive.

With regard to the properties of materials, e.g. considering cements, Neville (1994) warns of the dangers of assuming that the values of the properties of the cement actually used will be those shown in the cement test certificates because the certificates relate to bulk average quantities and not to individual consignments or to parts of consignments. To assist in minimising this quite normal problem, some cement works very helpfully provide pre-tested samples of cement for trials. In other cases major concrete producers with central testing facilities will test the cement used by their satellite laboratories for concrete testing. This applied to some of the data made available to the author during the trials but most were subject to the potential for error. The same problems apply to the other components of concrete, i.e. additions, admixtures and aggregates.

Additionally, the use of sieving to assess size distribution and mean size introduces the potential for inaccuracy because the process does not discriminate size other than relatively insensitively by boundary sieve sizes.

To overcome this problem for the aggregate mixture simulations, adjustments to size ratio were made by simple trial and error on the basis of observed differences based on theory.

For the concrete simulations, adjustments to match water demands from laboratory trials were assisted by utilising the Solver technique provided with Microsoft Excel. Solver incorporates a generalised reducing gradient algorithm developed by Lasdon *et al.* (1978) for optimising non-linear problems. This technique enables automatic adjustments to be made progressively to input values to reduce the differences between theoretical and actual output values to an acceptable level. The method may also assist in identifying the most probable source or sources of the differences. It is essential to consider the probability of such corrections being valid before finally accepting the full amount or some suitable proportion of the full corrections.

The following specific technique was developed for this purpose

- Possible sources of variation are selected.
- The **differences** are calculated between the observed and simulated property at a number of points over the range of data.
- The **mean difference** is calculated.[1]
- The **target mean difference** to be attained by the use of Solver is set at zero or other low value.
- Solver is set to operate on the input data. When the operation is complete the adjusted input data and new simulated output data are examined. Where adjustments are seen to be insignificant they can be ignored leaving only significant changes or some appropriate fraction of them to be adopted.[2]

Table B1 Situation before applying Solver to adjustment of water demand for Concrete Series 1 – Material code K

Laboratory trials		Simulation		Free water	
Cement (kg/m^3)	Free water (l/m^3)	Cement (kg/m^3)	Free water (l/m^3)	Difference (l/m^3)	Sq. diff.
100	190	100	199	9	79
200	170	200	179	9	90
300	162	300	174	12	151
400	166	400	179	13	170
500	180	500	196	16	263
			Sum	60	753
			Average	12	151

Example

Tables B1–B3 illustrate the situations before and after applying Solver to assess and adjust input data to reduce the differences between observed and simulated water demands of concrete for Concrete Series 1, Material code **K**. Consideration of the original materials test data suggested that the differences in Table B1 were most probably due to the cement properties differing between the certificated data for the bulk material and the cement sample used in the concrete. In particular, the marked increase in the difference with increasing cement content suggested that cement properties were a key aspect for investigation.

It will be seen from Table B3 that the original substantial bias and skew have been substantially reduced.

Table B2 Situation after applying Solver to adjustment of water demand for Concrete Series 1 – Material code K

Laboratory trials		Simulation		Free water	
Cement (kg/m^3)	Free water (l/m^3)	Cement (kg/m^3)	Free water (l/m^3)	Difference (l/m^3)	Sq. diff.
100	190	100	194	4	19
200	170	200	171	1	1
300	162	300	160	−2	3
400	166	400	162	−4	15
500	180	500	180	0	0
			Sum	0	37
			Average	0	7

Table B3 Situation after final compromise adjustment to input data for determining water demand for Concrete Series 1 – Material code K

Laboratory trials		Simulation		Free water	
Cement (kg/m^3)	Free water (l/m^3)	Cement (kg/m^3)	Free water (l/m^3)	Difference (l/m^3)	Sq. diff.
100	190	100	196	6	36
200	170	200	174	4	17
300	162	300	165	3	8
400	166	400	166	0	0
500	180	500	181	1	1
			Sum	14	62
			Average	3	12

Table B4 Adjustments made to input data for cement for determining water
demand for Concrete Series 1 – Material code K

Input data			
Property	*Before adjustm't*	*After initial adjustm't*	*After final adjustm't*
Cement			
Mean size	0.015	0.009	0.012
Voids ratio	0.9	0.81	0.81

Table B5 Measured and adjusted materials data for Concrete Series 2.

Data code	Materials	Properties				
		Mean size (mm) D	*Adj*	*Voids ratio U*	*Adj*	*Rel densy ssd RD*
P1	Cement	0.013		0.955	0.9	3.2
	Fine aggregate	0.595		0.66	0.6	2.60
	Coarse aggregate	10.71		0.8		2.55
P3	Cement	0.013		0.87	0.81	3.2
	Fine aggregate	0.61		0.67	0.58	2.62
	Coarse aggregate	10.69		0.66		2.56
P4	Cement	0.014		0.87		3.2
	Fine aggregate	0.65	0.53	0.71	0.60	2.61
	Coarse aggregate	9.63	10.5	0.79	0.74	2.57
P5	Cement	0.013		0.86		3.2
	Fine aggregate	0.53		0.58	0.61	2.65
	Coarse aggregate	11.4		0.77		2.55

The changes made to the cement properties are summarised in Table B4.
The change in mean size from 0.015 mm to 0.009 mm suggested by the use of
Solver seemed unlikely and a compromise value of 0.012 mm was selected
finally.

Table B5 shows original materials data from Concrete Series 2 and adjusted
data utilising the same technique, as a result of which, variation has been
significantly reduced in water demand and plastic density. The adjusted data
have been used in the detailed analysis in section 6.3.2.

It will be observed that the void ratios of the fine aggregates were the
commonest properties requiring adjustment. Densities are less variable and
provided the tests are carried out properly should not require adjustment, as
applies here.

Appendix C   Relations between cement and concrete properties

The relevance of cement properties for determining concrete properties is well known, e.g. Abrams (1924) included the water demand of cement in a formula for assessing concrete water demand; Murdock (1960) recognised the significance of the Vicat test of standard consistence for water demand of concrete.

Powers (1968) identified that the relatively high voids ratio of cements and other fine powders was due to the finest particles introducing significant interparticle cohesion, whether in a dry condition, or the inundated condition occurring in cement paste and concrete.

More recently, Neville (1994) highlighted the need to continue to study and test cement and cement paste, such tests being necessary to give basic insights into concrete behaviour. Neville stressed the importance of the other components in concrete and the interactions which would not be present in tests of cement or cement paste alone. These are basic tenets of the approach of the author in this present work.

In this appendix, research publications provided to the author by the late Professor G. Wischers of the Verein Deutscher Zementwerke e.V., Dusseldorf, are examined together with other published and unpublished work to assess the validity of the proposed methods of utilising cement test data for modelling water demand and also the strength of concrete.

In particular, commentary is provided from workers who have explored ways of modifying cements with regard to reducing water demand.

C1   Research of Rendchen on cement and concrete properties

C1.1   *Relations between cement properties and concrete water demand*

Rendchen (1985) reported properties of 22 cements, from which it has been possible to estimate data considered to be of relevance for this present work. A summary is provided in Table C1.[1]

Rendchen (1985) also reported tests of concretes at two water/cement ratios with each of the cements.

Table C1 Properties of cements investigated by Rendchen (1985)

Cement code	Fineness (m^2/kg)	Mean size log basis (mm)	RD in kerosene	Assumed RD in water	Vicat water %	Void ratio
H1	271	0.0160	3.13	3.19	25.0	0.794
H2	293	0.0169	3.08	3.14	24.5	0.766
H3	300	0.0153	3.09	3.15	25.5	0.800
H4	308	0.0141	3.14	3.20	24.0	0.764
H5	321	0.0147	3.11	3.17	26.0	0.822
H6	282	0.0117	3.13	3.19	31.5	1.005
H7	344	0.0108	3.10	3.16	26.5	0.835
H8	361	0.0103	3.09	3.15	26.5	0.832
H9	373	0.0097	3.07	3.13	27.0	0.843
H10	371	0.0089	3.08	3.14	26.5	0.830
H11	535	0.0069	3.13	3.19	27.5	0.875
H12	374	0.0090	2.96	3.02	28.5	0.859
NH1	403	0.0104	3.10	3.16	28.0	0.883
NH2	487	0.0095	3.08	3.14	28.5	0.893
NH3	602	0.0065	3.07	3.13	31.5	0.986
NH4	437	0.0093	3.06	3.12	29.0	0.903
NH5	517	0.0073	3.07	3.13	33.5	1.049
NH6	418	0.0098	3.10	3.16	27.0	0.851
S1	347	0.0106	3.14	3.20	30.0	0.959
S2	523	0.0078	3.06	3.12	32.5	1.014
S3	341	0.0097	3.16	3.22	32.0	1.031
S4	510	0.0066	3.14	3.20	33.5	1.073

At a w/c of 0.45, Rendchen demonstrated strong positive correlations between concrete water demand and cement water demand for 21 out of the 22 cements and a weaker correlation with specific surface or R-R parameters (section 2.1.2). The positive correlation with specific surface could imply a negative correlation with mean size, which is apparently at variance with the proposed theory. This is considered again later. At a w/c of 0.6, no strong correlations were discernible.

Rendchen additionally used linear multiple regression techniques to obtain improved prediction for the concretes of 0.45 w/c, involving use of the following parameters for the cement

• Cement water demand (Vicat test)
• Blaine fineness
• Cement water demand × Blaine fineness
• (Blaine fineness)$^{0.5}$
• A selected psd parameter.

The properties of the cements considered relevant to the present theory, i.e. mean sizes and voids ratios, and the observed water demands of the concretes are summarised in Table C2 together with 'predictions' based on linear multiple regression analyses made by the author, the results of which are summarised in Table C3.

The formula used for 'prediction' of water demand is

$$W_{pred} = I + A_1 \times d_p + A_2 \times U_p \tag{C1}$$

where I, A_1 and A_2 are obtained from the regression analysis for the particular w/c and d_p and U_p are the mean size and void ratio of the cement.[2]

The overall R^2 value for the 0.45 w/c data is sufficiently high to support the contention of a relation between both voids ratio and mean size of cement and the water demand of concrete although the degree of correlation with mean size is relatively poor. The lower value of R^2 for the 0.60 w/c data is to be expected, because of the dilution of the effect of cement at a higher w/c

Table C2 Data on cement properties calculated from data of Rendchen (1985) together with a comparison of observed and 'predicted' water demands of concrete based on regression analysis for this present work

Cement code	Cement properties		Concrete properties				
			w/c = 0.45		w/c = 0.60		
	Mean size (mm)	Voids ratio	Obs water (kg/m³)	Pred water (kg/m³)	Obs water (kg/m³)	Pred water (kg/m³)
H1	0.0160	0.794	159	162	150	154
H2	0.0169	0.766	158	158	155	153
H3	0.0153	0.800	164	162	154	154
H4	0.0141	0.764	153	157	149	151
H5	0.0147	0.822	163	165	152	154
H6	0.0117	1.005	207	188	174	160
H7	0.0108	0.835	169	165	155	152
H8	0.0103	0.832	162	164	150	151
H9	0.0097	0.843	169	165	156	151
H10	0.0089	0.830	158	163	144	150
H11	0.0069	0.875	173	168	150	150
H12	0.0090	0.859	181	167	159	151
NH1	0.0104	0.883	171	171	153	153
NH2	0.0095	0.893	167	172	151	153
NH3	0.0065	0.986	185	183	165	155
NH4	0.0093	0.903	171	173	152	154
NH5	0.0073	1.049	185	192	156	159
NH6	0.0098	0.851	162	166	150	152
S1	0.0106	0.959	174	181	153	157
S2	0.0078	1.014	186	187	154	157
S3	0.0097	1.031	185	191	155	160
S4	0.0066	1.073	194	195	153	159

Table C3 Results of multiple linear regression for the 'prediction' of water demand
of concrete from cement properties using the data in Table C2

Multiple linear regression for predicted water content of concrete

	$w/c = 0.45$	$w/c = 0.60$
R^2	0.77	0.28
Intercept	47.55	105.48
A1	459.1	780.4
A2	134.4	45.16

value, and to masking of effects by normal testing error both for cement and
for concrete; it is also consistent with the findings of Rendchen.

The positive signs for constants A_1 and A_2 imply that **increasing** mean size
and voids ratio are both to be associated with **increasing** water demand of
concrete. Thus, a finer size *per se* of cement does not lead directly to higher
water demand. However, if the increased fineness is accompanied by a steeper
psd, as will commonly be the case, then the voids ratio will be increased,
resulting in a higher water demand, as confirmed also by Bennett and Collings
(1969) and Sumner *et al.* (1989) as described in section C2.

For example, 10 of 12 cements in the H series of Rendchen demonstrate
increasing slope *n* with increasing fineness. Not surprisingly, Rendchen found
that in this situation, increasing water demand of concrete correlates with
increasing fineness. Thus, because of this auto-correlation it is also not
surprising to have obtained a relatively poor correlation with mean size for all
22 cements in the present analysis.

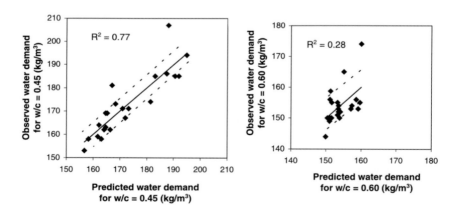

Figure C1 Comparison between the observed and 'predicted' water demands of
concrete based on the regression analysis utilising the cement properties.

A comparison is shown in Figure C1 between the 'predicted' and observed water demands to assist overall judgement of the analysis. The full lines shown are for the regression equations. The dotted lines are plotted at ± 5 kg/m^3 about the regression lines to demonstrate that the majority of the data are within a very narrow band. Only 2 of the 22 results for the concretes of 0.45 w/c are significantly outside the arbitrary band.

The data of Rendchen have also been utilised to test more directly the validity and predictive power of the Theory of Particle Mixtures. The results for water demand and strength and certain other properties are summarised in Table C4 and Table C5. The slump values are assumed. Also shown are the 'predicted' concrete properties based on the theories in this publication and the data for the cements in Table C1.

The overall aggregate grading used in the concrete tests conformed to midway A32-B32 in DIN 1045; thus the coarse aggregate had a maximum size of 32 mm. The mean sizes of the fine and coarse aggregates were calculated as 0.85 mm and 13.13 mm respectively. In the absence of information, it has been necessary to make assumptions concerning the voids ratios of the aggregates. Values of 0.525 and 0.5 for the fine and coarse aggregates respectively were selected as likely values for the described materials and yielded mean water demands close to those observed at the assumed slump of 50 mm.

The initial differences between the predicted and observed mean values for the examined properties are summarised in Table C6. The individual predictions were then adjusted to reduce the mean differences to zero. Table C4 and Table C5 contain **only** the adjusted values. No significance is attached to the differences which may be due to a variety of reasons, in particular the necessity to make assumptions about the aggregate voids ratios and the slump value.

Rendchen maintained a constant 35% for the fine to total aggregate irrespective of the cement properties and w/c value, whereas the Theory of Particle Mixtures varies the per cent fines automatically to maintain safe cohesion. Fortunately, the mean predicted per cent fines values, 29% and 37% for the 0.45 and 0.60 w/c concretes, are not very different from the value of 35% adopted by Rendchen. Thus, it is unlikely that the differences will have had a major effect on the comparison, although individual values may be affected.

The values for the statistical parameter, R^2, used to assess the degree of correlation from the regression analysis between the 'predicted' and observed water demands of the concretes are shown in Table C7 for the two w/c values separately and combined.

There is reasonable correlation between the 'predicted' and observed water demands of concretes made with the 22 cements, as may be seen from Table C7 and Figure C2, providing confidence that the theoretical assumptions are valid.

Table C4 Observed and 'predicted' properties of concrete for concretes of 0.45 w/c for the 22 cements examined by Rendchen (1985)

Cement code	Cement properties				Concrete observed properties						Concrete predicted properties			
	Mean size mm	Void ratio	Relative density	28d Str N/mm²	Cement kg/m³	water kg/m³	Compaction index v	Approx slump mm	Air %	28d Str N/mm²	Slump mm	Cement kg/m³	Water kg/m³	28d Str N/mm²
H1	0.0160	0.794	3.19	42	353	159	1.160	50	1.3	54	50	359	163	42
H2	0.0169	0.766	3.14	48	351	158	1.170	50	1.4	60	50	351	160	48
H3	0.0153	0.800	3.15	52	365	164	1.170	50	1.6	56	50	351	160	52
H4	0.0141	0.764	3.20	52	341	153	1.170	50	1.0	60	50	353	161	54
H5	0.0147	0.822	3.17	47	363	163	1.140	50	1.0	58	50	373	170	48
H6	0.0117	1.005	3.19	60	460	207	1.140	50	1.0	52	50	430	194	63
H7	0.0108	0.835	3.16	60	375	169	1.150	50	1.0	64	50	368	166	63
H8	0.0103	0.832	3.15	62	360	162	1.150	50	1.0	61	50	366	165	65
H9	0.0097	0.843	3.13	58	375	169	1.150	50	1.1	64	50	367	165	61
H10	0.0089	0.830	3.14	62	350	158	1.150	50	1.1	64	50	346	157	64
H11	0.0069	0.875	3.19	69	385	173	1.140	50	1.0	70	50	350	158	73
H12	0.0090	0.859	3.02	52	380	181	1.140	50	1.2	51	50	376	168	54
NH1	0.0104	0.883	3.16	60	380	171	1.140	50	1.4	60	50	373	168	62
NH2	0.0095	0.893	3.14	58	372	167	1.140	50	1.0	61	50	390	174	61
NH3	0.0065	0.986	3.13	58	410	185	1.140	50	0.9	66	50	408	183	61
NH4	0.0093	0.903	3.12	60	380	171	1.150	50	1.1	62	50	390	174	63
NH5	0.0073	1.049	3.13	64	410	185	1.150	50	1.1	63	50	420	191	67
NH6	0.0098	0.851	3.16	56	359	162	1.170	50	1.0	65	50	373	167	59
S1	0.0106	0.959	3.20	60	387	174	1.150	50	1.3	59	50	403	180	63
S2	0.0078	1.014	3.12	65	413	186	1.140	50	1.1	64	50	415	188	68
S3	0.0097	1.031	3.22	72	410	185	1.150	50	1.2	62	50	423	190	76
S4	0.0066	1.073	3.20	71	430	194	1.150	50	1.2	65	50	418	189	74
Average				58.5	382.2	172.5	1.2	50.0	1.1	61.0	50.0	382.0	172.5	61.0

Table C5 Observed and 'predicted' properties of concrete for concretes of 0.60 w/c for the 22 cements examined by Rendchen (1985)

Cement code	Cement properties				Concrete observed properties						Concrete predicted properties			
	Mean size mm	Void ratio	Relative density	28d Str N/mm^2	Cement kg/m^3	Water kg/m^3	Compaction index v	Assumed slump mm	Air %	28d Str N/mm^2	Slump mm	Cement kg/m^3	Water kg/m^3	28d Str N/mm^2
H1	0.0160	0.794	3.19	42	250	150	1.140	50	1.0	42	50	258	155	37
H2	0.0169	0.766	3.14	48	258	155	1.150	50	0.9	48	50	259	156	42
H3	0.0153	0.800	3.15	52	257	154	1.150	50	1.3	43	50	253	152	45
H4	0.0141	0.764	3.20	52	249	149	1.150	50	1.1	49	50	250	150	45
H5	0.0147	0.822	3.17	47	253	152	1.160	50	0.9	43	50	261	156	41
H6	0.0117	1.005	3.19	60	290	174	1.140	50	0.9	49	50	280	167	52
H7	0.0108	0.835	3.16	60	258	155	1.160	50	1.1	52	50	249	150	51
H8	0.0103	0.832	3.15	62	250	150	1.150	50	0.8	50	50	253	152	53
H9	0.0097	0.843	3.13	58	260	156	1.140	50	0.9	53	50	250	150	50
H10	0.0089	0.830	3.14	62	240	144	1.170	50	1.1	51	50	244	146	53
H11	0.0069	0.875	3.19	69	250	150	1.160	50	1.1	60	50	242	145	58
H12	0.0090	0.859	3.02	52	265	159	1.160	50	1.3	36	50	244	146	45
NH1	0.0104	0.883	3.16	60	255	153	1.150	50	1.6	48	50	245	147	50
NH2	0.0095	0.893	3.14	58	251	151	1.140	50	1.3	52	50	248	149	49
NH3	0.0065	0.986	3.13	58	275	165	1.130	50	0.7	53	50	266	159	50
NH4	0.0093	0.903	3.12	60	253	152	1.160	50	1.0	51	50	256	154	52
NH5	0.0073	1.049	3.13	64	260	156	1.160	50	1.2	53	50	269	161	54
NH6	0.0098	0.851	3.16	56	250	150	1.150	50	1.2	51	50	247	148	48
S1	0.0106	0.959	3.20	60	255	153	1.150	50	1.5	48	50	256	154	50
S2	0.0078	1.014	3.12	65	257	154	?1.16	50	0.9	59	50	271	163	56
S3	0.0097	1.031	3.22	72	258	155	1.160	50	1.0	50	50	275	165	61
S4	0.0066	1.073	3.20	71	255	153	1.160	50	0.9	60	50	274	165	60
Average					256.8	154.1	1.2	50.0	1.1	50.0	50.0	256.8	154.1	50.1

Table C6 Mean differences between 'predicted' and observed properties of concrete in Table C4 and Table C5 **before** adjustment

Parameter		Mean difference – Predicted – Observed	
		$w/c = 0.45$	$w/c = 0.60$
Cement	kg/m^3	13	15
Water	kg/m^3	3.7	8.7
Strength	N/mm^2	3.8	−4

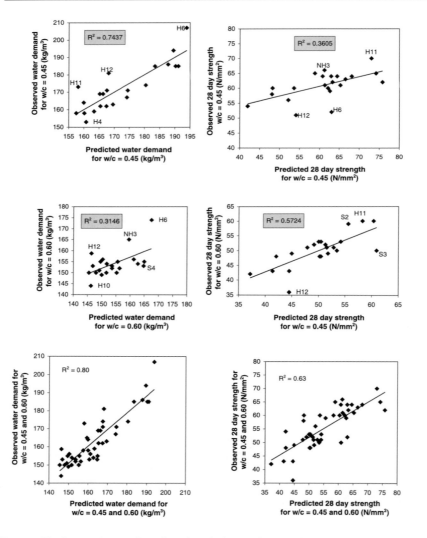

Figure C2 Comparisons of predicted and observed water demands and strengths for two water/cement ratios considered separately and together.

Table C7 Correlation coefficients from the regression analysis of the 'predicted' and observed values for water demands and strengths of concrete

	Coefficient R^2	
	---	---
w/c	Water content	Strength
0.45	0.74	0.31
0.6	0.36	0.57
Combined	0.80	0.63

C1.2 *Relations between cement strength and concrete strength*

The extension of the Theory of Particle Mixtures to strength is covered in section 5.4.2 and assumes a relationship between cement strength and concrete strength. The ability of the Theory of Particle Mixtures to predict strength of concrete from cement properties can be judged to be reasonably valid from examination of Table C7 and Figure C2.

C2 Research of Bennett, Sumner, Sprung, Krell and others

Additional confirmation of the validity of the present approach is provided by examination of data from Bennett and Collings (1969) for 4 Portland cements, including a high fineness cement S, and also a high alumina cement H, in concretes having cement contents of approx. 500 kg/m³. Again, it was necessary to make assumptions concerning the properties of the aggregates used in the concretes and to adjust for slump. Comparisons of observed and 'predicted' water demands based on the Theory of Particle Mixtures are shown in Table C8 after small adjustments to reduce the mean difference to zero.

Sumner *et al.* (1989), in a highly significant paper, concluded that

1. The higher paste water demand for standard consistency associated with a narrow particle size distribution is attributed to a higher voidage.
2. The higher concrete water demand seen for narrow particle size distribution cements is partly attributed to the physical influence of the particle size distribution and lower SSA but mainly to the lower level of gypsum dehydration which results in insufficient supply of available soluble SO_3.
3. The concrete water demand penalty associated with narrow particle size distribution cements can be reduced or even eliminated by ensuring an adequate level of dehydrated gypsum in the final cement.

Table C8 Comparison of observed and 'predicted' water demands of concrete for data obtained by Bennett (1969)

Cement code	SS (m^2/kg)	Vicat test water (%)	Est'd Mean size (mm)	Void ratio	RD asumed	Concrete Water Predicted	(kg/m^3) Observed
O	277	25	0.0162	0.827	3.2	193	185
R	490	28.5	0.0092	0.941	3.2	202	200
S	742	39.5	0.0061	1.298	3.2	245	260
C	489	29	0.0092	0.957	3.2	204	200
H	361	21.8	0.0125	0.752	3.33	177	175
					Average	204	204

Sumner observed that the finer cements and also ground slags with steep psds, i.e. narrow size distributions, had higher voidages and higher paste water demands than those with wider distributions.

Sumner also noted that concrete water demands were much less affected than paste water demands, due to a dilution effect of aggregate in concrete. Sumner attributes direct water reduction of finer cements as being

> a result of increased surface area (SSA) improving water retention properties of the powder thus enhancing lubrication of aggregate particles.

Sumner does not appear to have considered the possibility of an additional reduction in voidage and thus also water demand, due to reduced particle interference in concrete associated with the finer mean size. This reduction in particle interference would not apply in pastes due to the absence of fine aggregate.

Additionally, Sumner refers to considerations of gypsum solubility but it is not made clear why pastes and concretes would be affected differently unless time of test, which is not reported, was also a variable. Indeed, Sumner states that

> This does not help to understand the higher concrete water demands for narrow granulometry cements.

Sumner concludes that the

> concrete water demand penalty of using finer cements can be reduced or even eliminated by ensuring an adequate level of dehydrated gypsum in the final cement.

Thus, as suggested in section 5.3 chemical influences may also have significant effects on water demand which need to be taken into account.

Sprung (1986) discusses the various interactions between 'position parameter' and 'slope' of the Rosin-Rammler distribution and the influences of cement grinding techniques. Sprung observed that the benefits for water demand of lower position parameter (i.e. lower mean size) are usually offset by steeper or narrower particle size distributions.

Sprung also considered that fly-ash with a large surface area and favourable distribution can improve the rheological properties and structure of cement paste and concrete. Sprung noted that intergrinding of limestone with cement clinker leads to a reduction in water demand due to the preferential grinding of the limestone producing a broader size distribution with consequent improved void filling.

Krell (1985) examined a range of cements and tested them in concrete, examining in particular the effects on workability or consistence at constant water/cement ratio and cement content. The results of Krell show increasing water demand of cement as the psd narrows, with consequent reduction in spread in the flow test at constant w/c and cement content, as would be expected from the present work.

Krell identified the importance of accounting for, and distinguishing between, water held between particles and the water film held or associated with the surfaces of particles. Thus, for very fine materials, e.g. cement, silica fume, silt in aggregates etc., the water demands may be significantly greater than might be expected from considerations of particle size distribution and shape alone. Sprung (1986) also identified that cement paste water demand **increases** as the cement size is reduced while maintaining a constant slope of the psd.

Thus, it is important that the water demands of powders are assessed, either by taking account of such effects as suggested by Krell, or using test methods that will assess the total effects, e.g. Vicat test, as adopted by the author.

Also, with very fine materials, even when tested in water, voids ratios may be found to be significantly higher than expected because of high physical forces causing agglomeration of particles. as is discussed under section C3. Thus, it is important that the test method for water demand employs an energy level comparable with that to be employed for the concrete in the laboratory and in production. It may also be necessary in materials preparation or in concrete production to overcome the agglomeration, e.g. by use of pre-blending in a water/powder slurry; high energy mixing; vibration; incorporation of a plasticiser. Some of these have already been employed successfully in practice to increase the efficient use of expensive additions such as silica fume.

C3 Additional supporting information

Bache (1981), associated surface forces with the relatively open structure of cement paste, suggesting that these could be offset by the use of super-plasticisers, and could be benefited further by adjustment of the particle size distribution of the cement.

Hope (1990) concluded that various chemical factors and cement grading, but not fineness, had significant influences on water demand of concrete.

Odler and Chen (1990) reported that intensive grinding to produce finer cements resulted in a higher water demand due to a combination of reasons, including the narrower psd and various chemical factors. Kataoka *et al.* (1984) and Uchikawa (*c.* 1993) recognised the necessity for a wide size distribution for cement to minimise water demand which is echoed by Schnatz *et al.* (1995) who identified that a narrow size distribution of cement resulted in a high water demand

Bensted (1992) described the development and uses for microfine cements, i.e. cements having a mean size of say 5 μm (0.005 mm) and a specific surface of 500–1000 m²/kg or higher. Microfine cements have particular virtues for use in grouting for oilwell and construction purposes. However, Bensted identified the disadvantage of the high water demand of microfine cements requiring the use of super plasticisers. For the grouting applications described, widening the psd to reduce the water demand would be unacceptable because of the necessary low maximum size needed for grouting. For other potential applications, blending with coarser cementitious components could enable benefits to be obtained.

Reinhardt (1993) noted that cement suspensions may require special additions if agglomeration is to be prevented. Uchikawa (*c.* 1993) identified that the benefit of the very fine size of silica fume was offset by agglomeration but that this could be reduced by use of another fine material, e.g. finely ground slag, or limestone of intermediate size between that of cement and silica fume.

Uchikawa (*c.* 1993) refers to the significant benefit for workability and water demand of using cement made with spherical particles, implying blending rather then intergrinding to avoid losing the shape benefit.

It is important to distinguish intergrinding from blending. For example, Lange and Mortel (1995) reported that intergrinding of different materials usually results in particle size distributions of the individual materials which are rather haphazardly matched to one another due to large differences in hardness.

Yang and Jennings (1995) identified the importance of the mixing method in breaking down agglomeration of cement particles. Yang suggested that in field concrete, the cement paste is subject to ball-milling by aggregates.

Moir (1996) concluded that grinding techniques rather than other manufacturing aspects have the greatest potential to change cement characteristics, in particular to produce a more uniform finer size but that such a material may have an increased water demand due to the steeper grading. The inclusion of finely ground materials such as limestone can overcome this tendency.

Wang (1997) demonstrated that in considering the optimisation of hardened concrete, a wide psd was necessary to provide a high packing

Table C9 Analysis of cement and concrete data from C&CA for c. 1960

Cement			Concrete			Analysis				
				Observed free water at $c = 545$ kg/m^3						
Cement code	Fineness Blaine m^2/kg	Vicat water dmd %	Slump mm	At observed slump kg/m^3	Adjusted to 50 mm slump kg/m^3	uc	dc	1/dc	Theor d(W)	Obs d(W)
1	330	24	65	188	184	0.764	0.0167	60	−9	−6
2	430	31	30	188	196	0.992	0.0128	78	4	5
3	320	24	85	188	180	0.764	0.0172	58	−9	−10
4	440	29	45	188	190	0.927	0.0125	80	5	−1
5	260	26	55	188	187	0.829	0.0212	47	−3	−4
6	450	32	40	188	191	1.024	0.0122	82	8	1
7	350	28	40	188	191	0.894	0.0157	64	−6	1
8	410	28	40	188	191	0.894	0.0134	75	0	1
9	380	27.5	65	188	184	0.878	0.0145	69	−4	−6
10	550	30	0	188	219	0.959	0.0100	100	26	29
11	330	29	75	188	182	0.927	0.0167	60	−6	−9
12	370	27	40	188	191	0.862	0.0149	67	−5	1
13	440	27.5	50	188	188	0.878	0.0125	80	4	−2
14	300	27	85	188	180	0.862	0.0183	55	−7	−10
15	430	32	25	188	198	1.024	0.0128	78	5	8
16	310	27.5	65	188	184	0.878	0.0177	56	−7	−6
17	270	29	50	188	188	0.927	0.0204	49	−3	−2
18	450	31	20	188	201	0.992	0.0122	82	7	11
Average	379	28			190	0.904	0.0151	69	0	0

Table C10 Results of multiple linear regression for the 'prediction' of water
demand of concrete from cement properties using the data in Table C9

Multiple linear regression for predicted water content of concrete	
R^2	0.82
Intercept	−230
A1	17.37
A2	6038
A3	1.79

density but that a narrow [presumably fine] distribution was needed to obtain
a greater degree of hydration and thus lower porosity in the hardened
concrete. There was thus benefit in exploring the interaction of these opposing
concepts.

C4 Data analysis of cement and concrete data from C&CA from the 1960s

Unpublished data, made available to the author by the Cement and Concrete
Association (now British Cement Association), concerning cements and
concretes from the 1960s showed relatively poor correlation between
predicted and observed water demands of concrete when assessed using the

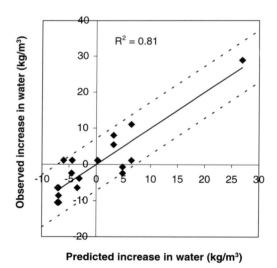

Figure C3 Predicted v observed increases in water demands of concrete for C&CA
data *c.* 1960.

theory but the inclusion of $1/d_c$ as a variable along with d_c and U_c in the multiple regression analysis improved the coefficient R^2 significantly.[3]

The original data and the results of the analyses are shown in Table C9. The multiple regression factors are shown in Table C10 and a comparison between 'predicted' and observed increases in concrete water demands, relative to the average for all cements, is shown in Figure C3.

The formula used for 'prediction' of the increase in water demand is

$$W_{pred} = I + A_1 \times U_c + A_2 \times d_c + A_3 \times 1/d_c \qquad (C2)$$

where I, A_1, A_2 and A_3 are obtained from Table C10 for the regression analysis and d_c and U_c are the mean size and voids ratio of the cement.[4]

If the correlation with $1/d_c$ is valid, it might be expected that inclusion of this parameter in the analysis of the Rendchen data discussed in C1.1 would result in an improved correlation, but this did not occur. Thus, the improved correlation obtained for the 1960 data may be fortuitous and is placed on record, but the parameter $1/d_c$ has not been included in the final selected modelling procedure to account for cement properties.

C5 Overview in respect of selected cement properties and their relationship to concrete properties

The examination of published and unpublished data confirm that the selected properties of

- Mean size of cement on a logarithmic basis and
- Voids ratio estimated from the cement water demand from the Vicat test

provide an adequate correlation with concrete water demand for the purposes of accurate modelling of concrete within the Theory of Particle Mixtures.

There is a reasonable correlation between EN450 cement strengths at 28 days and concrete strengths at 28 days made with particular aggregates.

Appendix D   Comparisons between different models for simulating aggregate combinations

In this appendix, comparisons are made between the author's model and, three models used by a number of researchers, namely

- The Aim and modified Toufar models discussed by Goltermann *et al.* (1997)
- The modified Mooney viscosity or 'solid suspension' model of de Larrard and co-workers

In Figure D1, for the Aim model referred to by Goltermann *et al.* (1997), the voids ratio diagrams are plotted for 4 sets of data from Series JDD1. The left-hand arm of the diagram is the theoretical line assuming no particle interference. The right-hand arm has an intercept of $0.90r$ on the left-hand axis. The Aim model, in effect, allows for particle interference only in the 'E–F zone' of the diagram and shows very poor simulation elsewhere, except for the lowest value of r illustrated in Figure D1, which accords with the conclusion of Goltermann *et al.* (1997). The value of n for minimum voids ratio and the value of U for minimum voids ratio would both be significantly under-estimated by using this model.

Goltermann *et al.* (1997) also examined the Toufar model and determined that it needed modifying slightly for low values of n. The modified model is examined in Figure D2. It will be seen that, for intermediate values of r, this model agrees well with the data and with the Dewar model, but that it under-estimated voids ratio at high r values and over-estimated at low r values. Goltermann's own experimental data show similar disagreement, suggesting that the Toufar model would need more radical modifications to be acceptable for general purposes when a high order of accuracy is sought.

The model adopted by de Larrard and co-workers is compared with the author's model in Figure D3. The experimental data are those of de Larrard for **high** energy compaction de Larrard and Buil (1987) and de Larrard and Sedran (1994) and Sedran *et al.* (1994). There is reasonable

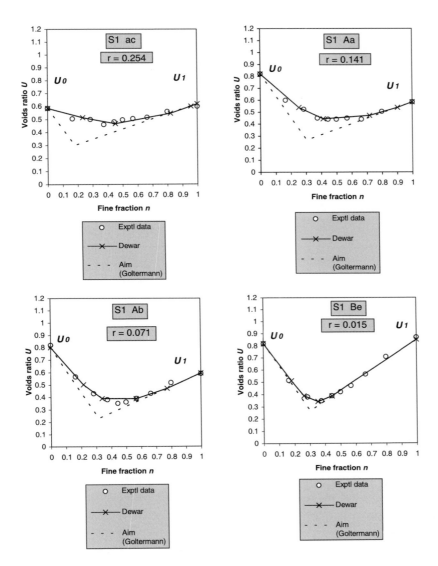

Figure D1 Assessment of the Aim model for predicting the voids ratio diagram of aggregate mixtures.

agreement between the diagrams based on the de Larrard and Dewar models.[1]

Of greater relevance to this present work is a comparison under **low** energy compaction. For this purpose, data provided by Sedran *et al.* (1994) have been used in Figure D4.[2]

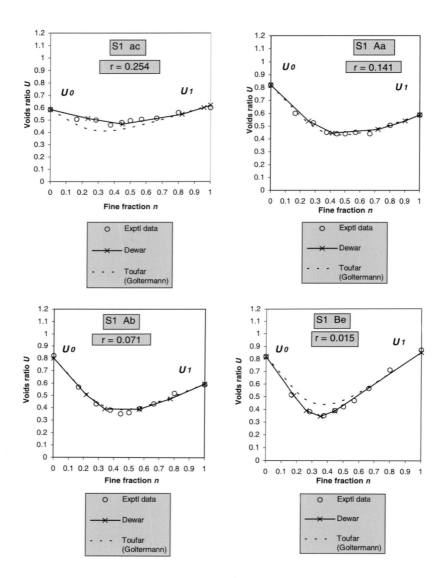

Figure D2 Assessment of the modified Toufar model for predicting the voids ratio diagram of aggregate mixtures.

It will be seen from Figure D4 that, whereas the Dewar model follows the de Larrard data accurately, the de Larrard model underestimated voids ratio at the higher r value examined and overestimated voids ratio at the lower r value.

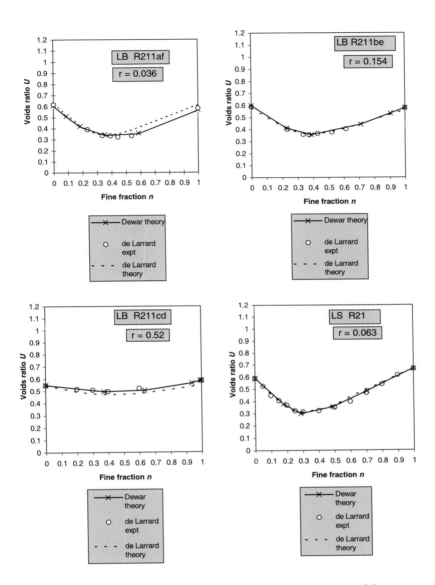

Figure D3 Assessment of the de Larrard *et al.* solid suspension model
for predicting the voids ratio diagram of aggregate mixtures under high
energy compaction.

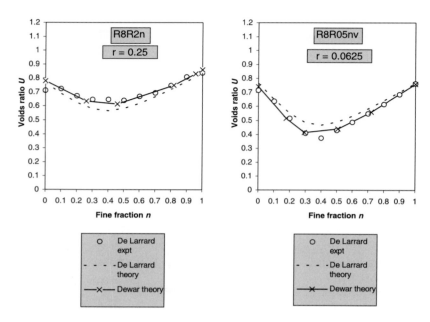

Figure D4 Assessment of the Sedran/de Larrard 'solid suspension' model for
predicting the voids ratio diagram of aggregate mixtures under low
energy compaction.

D1 Overview of alternative models

Of the alternative models examined, viz. Aim, modified Toufar or solid-
suspension of Sedran/de Larrard, only the Sedran/de Larrard model
compared favourably for accuracy against the author's model under **high**
energy compaction, but the Sedran/de Larrard model would need to be
modified further to perform accurately under the **low** energy compaction
method recommended in Appendix A.

Notes

2 The principal properties and test methods

1   For approximate assessment, the mean size is commonly close to the size corresponding to 50% passing in the particle size distribution.
2   Cements coded H, NH and S are respectively normal commercial cements, more finely ground commercial cements and specially prepared cements with modified gypsum contents. H1 to H11 are Portland cements, H12 is a Portland blastfurnace cement.
3   The relative density of water is assumed to be unity although the actual value will be dependent on ions in solution.
4   It is possible that the finest sizes of sands when tested separately should also be tested for voids ratio in water rather than in air. This was not done in the present work by the author but is recommended to be considered for similar work in the future.
5   In dry mixtures displacement of air is to be expected. In wet mixtures segregation of air and/or water needs to be considered.
6   If the aggregates have been tested for bulk density in a saturated and surface-dried condition, the relative density is that determined on an SSD basis. Alternatively, if the aggregates have been tested for bulk density in an oven-dried condition, the relative density on an oven-dried basis should be used.

3 Theory of Particle Mixtures

1   Powers used the term 'specific void content' in his earlier work but adopted 'voids ratio' latterly.
2   They have a particular advantage over other forms of diagram, e.g. void content diagrams which relate to the volume of the container, rather than to the volume of solids, because the essential theoretical relationships are straight lines.
3   Powers (1968) defined (n) as the solid volume of *coarse* material whereas the author uses **fine** material. Thus, the author's voids ratio diagrams of (u) vs. (n), in common with many other workers, are handed left to right compared with those of Powers.
4   From equation 3.10 when $n = 1$, then $U_n = -1$ irrespective of the value of U_0. Hence the point $(1, -1)$ is a fundamental point in all voids ratio diagrams.
5   Of course, individual results may transgress the boundaries due to normal experimental variation.
6   The dotted lines from U_0 and U_0'' converge at $(1, -1)$ below the diagram. For the example in Figure 3.6, point C is one such change point.

7   In Dewar (1983) the values of *m* were estimated from less data as 0.25, 1.0, 2.5 and 7.5 respectively at *B*, *C*, *D*, and *E* compared with 0.3, 0.75, 3 and 7.5 in the present more detailed analysis.

8   In earlier work, Dewar (1983) thought it necessary to apply different versions of the model in Figure 3.8 to take into account whether the mixture under consideration was, using Powers (1968) terms, coarse particle dominant or fine particle dominant. In this present work this distinction has been removed on the grounds that the same effects are present throughout the range of mixtures and can be accommodated within the same model.

9   Because of the influences of the various factors, the same arguments do not necessarily apply exactly to every example which might have been cited, and for which the descriptive terms chosen might be found to be less apt.

10   It is not to be expected that both methods will necessarily yield identical solutions.

4 Extension of the theory of particle mixtures to pastes, mortars and concretes

1   Normal concrete variation and the relatively few tests made with concretes may make it less likely that clear change points will be apparent with concretes, compared with mortars or aggregates. However, it is of interest to note that Ujhelyi (1997) identified three change points in the slope of relationships between water demand and cement content of concrete.

2   Values of mortar solids to coarse aggregate to the left of the lowest point produce concretes having a greater tendency to segregate whereas those to the right will tend to be over-cohesive, may lead to a mortar layer at the surface if over-compacted and have a slightly higher water demand.

3   A slightly more conservative value of 0.05 was adopted in Dewar (1983) compared with that now recommended of 0.025 as a start-point.

4   The reasons could be usefully investigated, together with an assessment as to whether the adjustments are influenced by the maximum size of coarse aggregate.

5   It will be observed in most cases that *d* and in 1 case *e* has the lowest voids ratio; this will vary with different materials.

6   In this example the sand is relatively coarse, resulting in substantial adjustments for some of the values of U at the change points.

5 Allowance for admixtures, air and other factors on water demand and strength of concrete

1   The cement used in the Vicat test should be of the same type and from the same source intended to be used in practice.

2   It was identified by Austin that the method and timing of adding the admixture to the water or to the paste in the Vicat test needed to follow closely the method and timing to be adopted for the concrete (see also section 5.3).

3   The entrained air contents have been assumed to be 1% less than the reported total air contents.

4   The data selected cover a relatively narrow range of cement strengths and for commonly available natural gravel aggregates and sands from East Anglia in use locally in the middle 1990s.

5   The effects of air and additions are dealt with later.

6   BCA (1990) and Dearlove (1991) provided relationships between EN 196 mortar prism and BS 4550 concrete cube strengths at different ages. The ratio of prism to cube strength at 28 days averages 1.30.

7   The corresponding range to that quoted is about 43 to 58 N/mm^2 in terms of the EN 196 mortar prism test.

8   There is also a corresponding effect observed by Hobbs (1976) that strength increases with aggregate content at constant w/c. The two effects may be connected because lower workabilities permit higher aggregate contents. This also accords with the conclusion of De Larrard and Belloc (1997) concerning the value of low paste thickness or coarse aggregate spacing.

9   The data for rhpc have been preferred to those for opc as probably being more closely related to today's ordinary Portland cements with regard to strength gain.

6 Comparisons between theoretical and experimental data for aggregates, mortar and concrete

1   The size chosen is smaller than that used normally for assessing coarse aggregates, but was chosen deliberately for ease of handling and it was assumed that the resulting slightly higher values of voids ratio for coarse aggregates, and mixtures containing coarse aggregates, would provide an additional safeguard through more cohesive mixtures for optimum conditions.

2   This applied only when the size ratio r was less than about 0.125, as for some of the reported results in Table 6.1 and Figure 6.1.

3   Size ratio is subject to experimental error resulting from normal variations in the sizes of materials used, compared with the nominal sizes assumed or calculated from the standard sieving test to BS 812. Included in Table 6.2, and other tables of aggregate test data, are adjustment factors applied to the size ratios to take account of residual discrepancies apparent in the analysis of the data and assumed to emanate from these variations (see also Appendix B).

4   Small adjustments, usually 0.03 or less, have been made to the values of *U0* or *U1* used in the equations compared with those observed in order to reduce residual discrepancies. These adjustments will be apparent to the reader when a theoretical value shown by a cross at $n = 0$ or 1 is not coincident with the experimental value.

5   Small adjustments were considered to be necessary for 19 of the 108 values for materials properties. In general, adjustments were made only when there was a strong probability of differences between the actual properties of materials and the sample test data or certificates.

6   The commonest case for such adjustments concerned the properties of the actual sample of cement compared with average certificated data applying to bulk consignments over a period. In some cases, this problem was avoided by testing the cement sample used for the concrete trials.

7   In appropriate cases of systematic deviations in a set of data, a computerised mathematical method of adjustment, developed by Lasdon *et al.* (1978), was used to minimise the deviations between observed and theoretical values. Its use is described in Appendix B.

8   It was not possible to ascribe reasons with certainty for the necessity for these adjustments. The general good agreement for the seven sets of data which did not require adjustment suggest that the theory does not require modification as a result of the need for adjustment for the other five. What is apparent is that for future research it is vital to ensure that the relative density values are valid and relate to the materials to be used and that the air content of the fresh concrete is measured.

7 Proportions and properties of concrete predicted by the use of the theory of particle mixtures

1   The gradings are in the centres of the permitted ranges of BS 882 for graded materials and the voids ratios have been calculated on the basis that the void ratio of **each size fraction** is 0.85. The method of calculation of the composite voids ratio involves successive combination of the smallest material with the next smallest and combining the mixture with the next size and so on, until all sizes have been incorporated. This method is described in more detail in section 8.4.
2   Effects may differ substantially for materials having different properties.

8 Case studies

1   In Table 8.1, column 4 assumes that the mean size of the addition is 7 µm compared with 15 µm for the cement. Column 6 assumes that the mean size of the addition is 15 µm, the same as that of the cement.
2   In Figure 8.1, the dotted line is the relationship that would be expected if the mean size of the addition was 7 µm. The full line is the relationship that would be expected if the mean size of the addition was 15 µm. In this latter case there would be no change in slope with proportion of addition.
3   Additional constraints can be applied in the optimisation process. In this example the maximum size was restricted to 0.08 mm (80 µm), the proportion passing 0.005 mm (5 µm) was restricted to a maximum of 15% by mass and it was assumed that no material passed 0.0025 mm (2.5 µm).
4   Additional constraints can be applied in the optimisation process. In this example the maximum size was restricted to 5 mm, the proportion passing 0.08 mm (80 µm) was restricted to a maximum of 4% by mass and it was assumed that no material passed 0.04 mm (40 µm).

9 A user-friendly computerised system for general application

1   Special sets of sieve sizes can be created or the reference set can be modified.
2   Dosages of Admixtures and their effects are entered via the Conditions screen.

Appendix B Adjustment of input data

1   If it seems more appropriate, squared differences and the root mean squared difference can be used.
2   If the use of Solver does not correct any marked skew in the initial output data, or if its use introduces a marked skew, it is probable that the properties selected for processing need reconsideration.

Appendix C Relations between cement and concrete properties

1   Cements coded H, NH and S are respectively normal commercial cements, more finely ground commercial cements and specially prepared cements with modified gypsum contents. H1 to H11 are Portland cements, H12 is a Portland blastfurnace cement.
2   The constants in this formula have applicability only to the particular results in Table C2 and differ between w/c values. Their purpose is limited to examining the contention that water demand of concrete is related to mean size and voids

ratio of cement. Based on the advice of Ehrenberg (1975), the term 'prediction' is used solely within the context of the particular regression analysis, and the formulae and constants are *not* to be considered as relevant or valid for general prediction purposes.

3   It will be noted that $1/d_c$ is related to specific surface and the implication is that the water demand of concrete increases as the specific surface of the cement decreases, at the same time as decreasing with mean size due to reduction in particle interference.

4   The constants in this formula have applicability only to the particular results in Table C9. Their purpose is limited to examining the contention that water demand of concrete is related to mean size and void ratio of cement. Based on the advice of Ehrenberg (1975), the term 'prediction' is used solely within the context of the particular regression analysis and the formulae and constants are *not* relevant or valid for general prediction purposes.

Appendix D Comparisons between different models for simulating aggregate combinations

1   The z values for **high** energy compaction tabled in Appendix A have been used in the Dewar model. However, as stressed in Appendix A, the author does not recommend **high** energy compaction for normal situations applying to aggregates and concrete.

2   The z values for **low** energy compaction have been used in the Dewar model. The Sedran/de Larrard model uses a different reference 'viscosity' for **low** energy compaction.

Bibliography

Abrams, D.A. Design of concrete mixtures. Bulletin 1. Structural Materials Research Laboratory. Lewis Institute, Chicago, July 1924.

ACI Standard Practice for selecting proportions for normal, heavyweight and mass concrete. ACI 211.1-89, pp 1–37.

ACI Guide for selecting proportions for high-strength concrete with Portland cement and fly ash. *ACI Materials Journal.* May–June 1993, pp 272–83.

ACI Guide to the use of silica fume in concrete. ACI 234R-96. 1996, pp 51.

Ackroyd, L.W. and Rhodes, F.G. An investigation of the crushing strengths of concrete made with three different cements in Nigeria. *Proceedings, ICE,* 1963, pp 325–40.

Al-Jarallah, M. and Tons, E. Void content prediction in two-size aggregate mixes. ASTM, *Journal of Testing and Evaluation.* Vol 9 No 1, 1981, pp 3–10.

Andersen, J. and Johansen, V. Computer-aided simulation of particle packing. A tool for proportioning cement based materials. GM Idorn Consult A/S, Denmark, *c.* 1989.

Austin, R. An investigation into how to incorporate the addition of water reducing admixtures into the Mixsim computer program. Project report. ACT Diploma 1994/5. Institute of Concrete Technology.

Bache, H.H. Densified cement/ultrafine particle-based materials. *2nd International Conference on Superplasticizers in Concrete,* June 1981, Ottawa, pp 1–35.

Ball, D. Selecting aggregates for maximum packing density in low permeability concretes. *Concrete,* May 1998, pp 9–10.

Banfill, P.F.G. and Carr, M.P. The properties of concrete made with very fine sand. *Concrete,* March 1987, pp. 11–14.

Barber, P. If you don't have 40 results, we will need site trials. *Concrete,* Sept/Oct 1995, pp 48–50.

Barber, P. Discussion of the factors that control the form of the cement content to strength relationship. *Proc. 12th ERMCO Congress,* June 1998, Lisbon. pp 322–8.

Barker, G.C. Computer simulations of granular materials. *Granular Matter – An Interdisciplinary Approach,* Springer-Verlag 1994, pp 35–83.

Baron, J., Bascoul, A., Escadeillas, G. and Chaudouard, G. From clinker to concrete – an overall modelling. *Materials and Structures,* No 26, 1993, pp 319–27.

BCA New test method for cement strength. Information Sheet No 2, November 1990, 2 pp.

Beaudoin, J.J., Feldman, R.F. and Tumidjaski, P.J. Pore structure of hardened Portland cement pastes and its influence on properties. *Advn Cem Bas Mat*, 1994, pp 224–36.

Beeby, A.W. Empiricism versus understanding, in the successful use of materials in a changing world. Editorial comment. *Magazine of Concrete Research*, Vol 43, No 156, September 1991, pp 141–2.

Bennett, E.W. and Collings, B.C. High early strength concrete by means of very fine Portland cement. ICE, Technical Note No 10, 1969, pp 443–52.

Bensted, J. Microfine cements. *World Cement*, Dec 1992, pp 45–7.

Bloem, D.L. and Gaynor, R.D. Effect of aggregate properties on strength of concrete. *Journal of the ACI*, October 1963, pp 1429–55.

Brookbanks, P. Properties of fresh concrete. BRE, Seminar, Performance of limestone filled cements, November 1989, pp 19.

Brown, B.V. Aggregate: the greater part of concrete. *Concrete 2000*, Dundee 1993, pp 279–93.

BSI Methods of testing cement. Part 3. Determination of setting time and soundness. BS EN 196-3: 1995, pp 1–10.

Butler, W.B. Superfine fly ash in high strength concrete. *Concrete 2000*, Dundee, 1993.

Chmielewski, T., Switonski, A. and Switonska, E. Computerised design of concrete for requisite properties. *Concrete 2000*, Dundee 1993, pp 1507–18.

Concrete Society. Micro silica in concrete. TR41, 1990.

Day, K.W. *Concrete Mix Design, Quality Control and Specification*. E & FN Spon, 1995, 350 pp.

Dearlove, A. New UK test method for cement strength. *Concrete*. May/June 1991, pp 50–2.

de Larrard, F. Modele lineaire de compacite des melanges granulaires. AFREM, Versailles, September 1987, pp 325–32.

de Larrard, F. and Buil, M. Granularite et compacite dans les materiaux de genie civil. *Materials and Structures*, 1987, Vol 20, pp 117–26.

de Larrard, F. and Tondat, P. Sur la contribution de la topologie du squelette granulaire à la résistance en compression du beton. *Materials and Structures*, 1993, Vol 26, pp 505–16.

de Larrard, F. and Sedran, T. Optimization of ultra-high-performance concrete by the use of a packing model. *Cement and Concrete Research*, 1994, Vol 24, No 6, pp 997–1009.

de Larrard, F. La résistance en compression des betons de structure aux cendres volantes. *Materials and Structures*, Vol 28, No 182, October 1995, pp 459–63.

de Larrard, F. and Belloc, A. The influence of aggregate on the compressive strength of normal and high strength concrete. *ACI Materials Journal*. V94, No 5, Sept/Oct 1997, pp 417–25.

de Schutter, G. and Taerwe, L. Random particle model for concrete based on Delaunay triangulation. *Materials and Structures*, No 26, 1993, pp 67–73.

Dewar, J.D. The workability of ready mixed concrete. *RILEM Symposium on Workability*, March 1973, Leeds.

Dewar, J.D. Computerised simulation of aggregate, mortar and concrete mixtures. *ERMCO Congress*, London, May 1983.

Dewar, J.D. Ready-mixed concrete mix design. *Mun. Engr*, 3, Feb, 1986a, pp 35–43.

Dewar, J.D. The structure of fresh concrete – A new solution to an old problem. First Sir Frederick Lea Memorial lecture, Institute of Concrete Technology, 1986b, pp 1–23. Reprinted by the BRMCA.

Dewar, J.D. and Anderson, R. *Manual of Ready-mixed Concrete*. Blackie; Chapman and Hall. First edition 1988; second edition 1992, pp 1–245.

Dewar, J.D. The application to cementitious systems of a general theory of particulate mixtures. *Cement and Concrete Science*. Conference of the Institute of Materials, Oxford. 26–27 September 1994, pp 1–8.

Dewar, J.D. A concrete laboratory in a computer. Case-studies of simulation of laboratory trial mixes. *ERMCO Congress*, Istanbul, 1995, pp 1–9.

Dewar, J.D. Development of a Theory of Particle Mixtures and its application to aggregates, mortars and concretes. PhD thesis, Dept of Civil Eng, City University, London, December 1997.

Dewar, J.D. Computer methods – Concrete proportioning. Inst. of Concrete Technology, *26th Annual Convention*, Market Bosworth, UK, April 1998, pp 1–14.

Dewar, J.D. Optimised design of concrete and its component materials. *Proc. 12th ERMCO Congress*, June 1998, Lisbon, pp 418–28.

Domone, P.L.J. and Soutsos, M.N. Properties of high-strength concrete mixes containing pfa and ggbs. *Mag. Concr. Res.*, Dec 1995, Vol 47, No 173, pp 355–67.

Drinkgern, G. Deviations from the 'classical' mix design for concrete. *Betonwerk und Fertigteil Technik*, 11/1994, pp 45–57.

Dunstan, M.R.H. Development of high fly ash content concrete. *Proc. ICE Part 1*, 1983, Vol 74, August, pp 495–513.

Ehrenberg, A.S.C. *Data reduction – Analysing and Interpreting Statistical Data*. John Wiley, 1975, pp 234–58.

Ehrenburg, D.O. An analytical approach to gap-graded concrete. *Cement, Concrete and Aggregates*, ASTM, Vol 2, No 1, Summer, 1980, pp 39–42.

Ehrenburg, D.O. Proportioning of coarse aggregate for conventionally and gap-graded concrete. *Cement, Concrete and Aggregates*, ASTM, Vol 3, No 1, Summer, 1981, pp 37–9.

El-Didamony, H., Amer, H.A., Ebied, E. and Heikal, M. The role of cement dust in some blended cements. *Il cemento* 4/1993, pp 221–30.

Erntroy, H.C. and Shacklock, B.W. Design of high-strength concrete mixes. Cement and Concrete Association. Reprint No 32, May 1954, pp 55–73 and 163–6.

Fagerlund, G. Economical use of cement in concrete. *Building Issues*, Vol 6, No 2, 1994, Lund University, Sweden, 24 pp.

Fainer, M. Sh. Concepts of optimum concrete design. Trans from *Beton i Zhelezobeton*, No 1, 1992, pp 15–16.

FIP *Condensed Silica Fume in Concrete*. Thomas Telford, 1988, 37 pp.

Foo, H.C. and Akhras, G. Expert systems and design of concrete mixtures. *Concrete International*, July 1993, pp 42–6.

Franklin, R.E. and King, T.M.J. Relations between compressive and indirect tensile strengths of concrete. DoE Road Research Laboratory, 1971.

Frohnsdorff, G. Clifton, J.R., Garboczi, E.J. and Bentz, D.P. Virtual cement and concrete. PCA, Emerging Technologies Symposium, March 1995.

Gaynor, R.D. High strength air-entrained concrete. NRMCA/NSGA, Joint research laboratory publication No 17, March 1968, pp 1–19.

Gaynor, R.D. and Meininger, R.C. Evaluating concrete sands. *Concrete International*, Dec 1983, pp 53–60.

Glanville, W.H., Collins, A.R. and Matthews, D.D. The grading of aggregates and workability of concrete. DSIR Road Research Laboratory, Road Research Technical Paper No 5, 1938, pp 1–38.

Glavind, M. and Jensen, B. Design of a stable air void system in concrete by optimization of the composition of the aggregate. *Radical Concrete Technology*. Spon 1996, 332–41.

Goltermann, P., Johansen, V. and Palbol, L. Packing of aggregates: an alternative tool to determine the optimal aggregate mix. *ACI Materials Journal*. V94, No 5, Sept/Oct 1997, pp 435–43.

Gray, W.A. *The Packing of Solid Particles*, Chapman & Hall, 1968, pp 1–34.

Gutierrez, P.A. and Canovas, M.F. High performance concrete: requirements for constituent materials and mix proportioning. *ACI Materials Journal*, May/June 1996, pp 233–41.

Hansen, T.C. Physical structure of hardened cement paste. A classical approach. *Materiaux et Constructions*, Vol 19, No 114, 1986.

Hansen, T.C. Modified DOE mix design method for high volume fly ash concretes and controlled low strength concretes. *Mag. Concr. Res.*, 1992, Vol 44, No 158, March, pp 39–45.

Hobbs, D.W. Influence of aggregate volume concentration upon the workability of concrete and ... *Mag. Concr. Res.*, Vol 28, No 97, December 1976. Discussion *Mag. Concr. Res.*, Vol 29, No 101, Dec 1977.

Hobbs, D.W. Mix design, water cement ratio, variability and quality of mixing water. *Seminar on Concrete and Water*, Saint Remy les Chevreuse, France, June 1985.

Hobbs, D.W. Portland-pulverized fuel ash concretes: water demand, 28 day strength, mix design and strength development. *Proc. ICE*, part 2, 1988, Vol 85, pp 317–31.

Hope, B.B. and Hewitt, P.M. Progressive concrete mix proportioning. *ACI Journal*, May–June 1985.

Hope, B.B. and Rose, K. Statistical analysis of the influence of different cements on the water demand for constant slump. Properties of fresh concrete. *Colloquium*, Hanover, 1990, pp 59–66.

Hughes, B.P. Rational Concrete mix design. *Proceedings ICE*, Vol 17, November 1960, pp 315–32.

Hughes, B.P. Rational concrete mix design. ICE November 1960, reprint by the British Granite and Whinstone Federation. Discussion *Proc ICE*, Vol 21, April 1962, pp 927–52.

Hughes, B.P. Mix design for local materials. *Advances in Ready Mixed Concrete Technology*, Pergamon Press, 1976, pp 253–68.

Jackson, P.J. Manufacturing aspects of limestone-filled cements. BRE, Seminar, *Performance of Limestone Filled Cements*, November 1989, 19 pp.

Johnston, C.D. Influence of aggregate void condition and particle size on the workability and water requirement of single-sized aggregate-paste mixtures. *Properties of Fresh Concrete. Colloquium*, Hanover, 1990, pp 67–76.

Kantha Rao, V.V.L. and Krishnamoothy, S. Aggregate mixtures for least-void content for use in polymer concrete. *Cement, Concrete and Aggregates*, Vol 15, No 2, Winter 1993, pp 99–107.

Kaplan, M.F. Effects of incomplete consolidation on compressive and flexural strength, ... of concrete. *ACI Journal*, Vol 56, March 1960, pp 853–67.

Kataoka, N., Maeo, M. and Koga, T. Properties of cement ground in a roller mill. *CAJ Review*, 1984, pp 84–7.

Kennedy, S.K. Fractal characterisation of concrete components. *Proc 15th Int Conf on Cement Microscopy*, March/April 1993, Dallas.

Kessler, H.-G. Spheres model for gap gradings of dense concretes. *Betonwerk und Fertigteil Technik*, 11/1994, pp 63–76.

Kirkham, R.H.H. Mix design for compressive and flexural strength using granite aggregate. British Granite and Whinstone Federation, March 1965. Pre-print.

Krell, J. Die Konsistenz von Zementleim, Mörtel und Beton und ihre zeitliche Veränderung. VDZ, Dusseldorf, Schriftenreihe fur der Zementindustrie, Heft 46, 1985, 117 pp.

Kronlof, A. Effect of very fine aggregate on concrete strength. *Materials and Structures*, 1994, 27, pp 15–25.

Lange, F. von and Mortel, H. The effect of additions of ultra-fine cement on the strength characteristic of mortars. *ZKG International*. No 12, 1995, pp 661–6.

Larsen, A. Partikelpackning-proportionering av betong. CBI, Stockholm, June 1991, pp 1–26.

Lasdon, L.S., Waren, A., Jain, A. and Ratner, M. Design and testing of a *generalised* reduced gradient code for non-linear programming. ACM Transactions on mathematical software, Vol 4, part 1, 1978, pp 34–50.

Lees, G. The rational design of aggregate gradings for dense asphaltic compositions. *Proc Assoc Asphalt Technology*, 39, 1970a, pp 60–97.

Lees, G. Studies of inter-particle void characteristics. *Q. Jl of Geology*, Vol 2, 1970b, pp 287–99.

Leutwyler, K. Shaking conventional wisdom. *Scient Amer*. Sept 1993, p 12.

Li, S.T. and Ramakrishnan, V. Discussion of 'Proportioning of coarse aggregate for conventionally and gap-graded concrete' by D.O. Ehrenburg. *Cement, Concrete and Aggregates*, ASTM, Vol 5, No 2, Winter, 1983, pp 145–6.

Loedolff, G.F. Private communication. Extracts from a PhD Thesis, University of Stellenbosch, 1985, pp 20.

Loedolff, G.F. Dense concrete and fly ash. CANMET conference on fly ash, slag, silica fume and pozzolans, Madrid, April 1986.

Marchand, J. *et al.* Mixture proportioning of roller compacted concrete – a review. ACI SP 171, 1997, pp 457–86.

McIntosh, J.D. *Concrete Mix Design*. CACA, 2nd edition, 1966.

Meyer, L.M. and Perenchio, W.F. Theory of concrete slump loss related to the use of chemical admixtures. *Concrete*, Part 1, February 1982, pp. 33–7 and Part 2, March 1982, pp. 33–6.

Moir, G.K. Cement production – state of the art. *ICT, 24th convention*, 1–3 April 1996, pp 1–37.

Moncrieff, D.S. The effect of grading and shape on the bulk density of concrete aggregates. *Mag. Concr. Res.* Dec 1953, pp 68–70.

Mooney, M. The viscosity of concentrated suspensions of spherical particles. *Journal of Colloids and Interface Science*, Vol. 6, 1951, p. 162.

Mortsell, E., Maage, M. and Smeplass, S. A particle-matrix model for prediction of workability of concrete. *Production Methods and Workability of Concrete*, Paisley, 1996, pp 429–38.

Mullick, A.K., Iyer, S.R. and Babu, K.H. Variability of cements and adjustments in concrete mixes by accelerated tests. Paper W14A(3), *ERMCO 83 Congress*, London, May 1983.

Murdock, L.J. The workability of concrete. *Mag. Conct. Res.*, Vol 12, No 36, Nov 1960, pp 135–44 and discussion, *Mag. Concr. Res.*, Vol 13, No 38, Jul 1961, pp 79–92.

Nagaraj, T.S., Shashiprakash, S.G. and Kameshwara, Rao. B. Generalised Abrams' law. *RILEM Colloquium, Properties of Fresh Concrete*, Hanover, 1990, pp 242–52.

Nagaraj, T.S. and Zahida, Banu. Generalisation of Abrams' law. *Cement and Concrete Research*, Vol 26, No 6, 1996, pp 933–42.

Nehdi, M., Mindess, S. and Aitcin, P-C. Optimization of triple-blended composite cements for making high-strength concrete. *World Cement Research and Development*, June 1996, pp 69–73.

Neville, A.M. Cement and Concrete: Their interrelation in practice. *Advances in Cement and Concrete*. ASCE, July 1994, pp 1–14.

Neville, A. and Aitcin P-C. High performance concrete – An overview. *RILEM, Materials and Structures*, Vol 31, March 1998, pp 111–17.

Newman, A.J. and Teychenne, D.C. A classification of natural sands and its use in concrete mix design. Mix design and quality control of concrete. Symposium, C&CA, London, May 1954, pp 175–207.

Newman, K. The effect of water absorption by aggregates on the water/cement ratio of concrete. *Mag. Concr. Res.* Vol 1, No 33, Nov 1959, pp 135–42. Discussion Vol 12, No 35, July 1960, pp 115–18.

Nielsen, L.F. Strength development in hardened cement paste: examination of some empirical equations. *Materials and Structures*, 1993, Vol 26, pp 255–60.

NRMCA. Outline and tables for proportioning normal weight concrete. April 1977, *NRMCA publication* No 154, 8 pp.

Numata, S. Constitutive rules for selecting mix proportions for concrete with the application of a new geometrical theory of particle interference between fine and coarse aggregates. July 1994, pp 17–26.

Numata, S. (Personal communication). Nishinippon Institute of Technology, 28 July 1995.

Odler, I. and Chen, Y. Influence of grinding in high-pressure grinding rolls on the properties of Portland cement. *Zement-Kalk-Gips* No 4, 1990, pp 188–91.

Oluokun, F.A. Fly ash concrete mix design and the water-cement ratio law. *ACI Materials Journal*, July/August 1994, pp 362–71.

Palbol, L. Optimization of concrete aggregates. *Betonwerk und Fertigteil Technik*, 11/1994, pp 58–62.

Persson, B. Seven year study on the effect of silica fume in concrete. *Adv. Cem. Bas. Mat.* Vol 7, Elsevier Science Ltd, 1998, pp 139–55.

Plum, N.M. The pre-determination of water requirement and optimum grading of concrete. *Danish Inst. for Bdg, Research. Study* No 3, Copenhagen, 1950, 96 pp.

Popovics, S. Some aspects of measuring consistence. *Mag. Concr. Res.*, Vol 17, No 50, Mar 1965, pp 15–20.

Popovics, S. *Concrete-making Materials*. Hemisphere Pub Co, 1979.

Popovics, S. New formulas for the prediction of the effect of porosity on concrete strength. *ACI Journal*, March/April 1985, pp 136–46.

Popovics, S. Analysis of the concrete strength versus water–cement ratio relationship. *ACI Journal*, September/October 1990, pp 517–29.

Popovics, S. The slump test is useless – or is it? *Concrete International*, September 1994, pp 30–3.

Popovics, S. and Popovics, J.S. The foundation of a computer program for the advanced utilisation of w/c and air content in concrete proportioning. *Concrete International*, December 1994, pp 26–6.

Powers, T.C. Some analytical aspects of fresh concrete. *Cement, Lime and Gravel*. Feb/March 1966, pp 29–36 and 67–73.

Powers, T.C. *The Properties of Fresh Concrete*. 1968, John Wiley, pp 1–664.

Previte, R.W. Concrete slump loss. *ACI Journal*, August 1977, pp 361–7.

Puntke, W. Mix design considerations for granulometric optimization of the matrix of high performance concrete. *Radical Concrete Technology*, Spon 1996, pp 129–40.

Reinhardt, H.W. Ultra-fine cements for special applications. *Advanced Cement Based Materials*, 1993, 1, pp 106–7.

Rendchen, K. Einfluß der Granulometrie von Zement auf die Eigenschaften des Frischbetons und auf das Festigkeits- und Verformensverhalten des Festbetons. VDZ, Dusseldorf, Schriftenreihe fur der Zementindustrie, Heft 45, 1985, 189 pp.

Roy, D.M. and Silsbee, M.R. Novel cement products for applications in the 21st century. *ACI Special Publication* SP-144, 1994, pp 349–82.

Roy, D.M., Scheetz, B.E., Malek, R.I.A. and Shi, D. *Concrete Components Packing Handbook*. Strategic Highway Research Program. NRC Washington DC. 1993, pp 1–161.

Schnatz, R., Ellerbrock, H-G. and Sprung, S. Influencing the workability characteristics of cement during finish grinding with high-pressure grinding rolls. *ZKG International*, Vol 48, No 5, 1995, pp 264–73.

Sear, L.K.A., Dews, J., Kite, B., Troy, J.F. and Harris, F.C. Abrams' law, air and high water cement ratios. *Construction and Building Materials*. Vol 10, No 3, 1996, pp 221–6.

Sedran, T., de Larrard, F. and Angot, D. Prevision de la compacite des melanges granulaires par le modele de suspension solide. Parts 1 and 2. *Bulletin liaison Labo. P. et Ch.* No 194, Nov/Dec 1994, pp 59–70 and pp 71–86.

Shilstone, J.M. Concrete mixture optimization. *Concrete International*, June 1990, pp 33–9.

Smith, I.A. The design of fly ash concretes. Paper 6982, *Proc ICE*, Vol 36, April 1967, pp 769–90.

Somerville, G. Cement and concrete as materials: changes in properties, production and performance. *Proc. ICE Structs and Bldgs*. No 116, Aug/Nov 1996, pp 335–43.

Sprung, S. Influence of process technology on cement properties. *Zement-Kalk-Gips* No 12, 1986, pp 309–16.

Stewart, D.A. *The Design and Placing of High Quality Concrete*. 2nd edn, Spon, 1962.

Sumner, M.S., Hepher, M.N. and Moir, G.K. The influence of a narrow cement particle size distribution on cement paste and concrete water demand. *Ciments, Betons, Platres, Chaux* No 778, March 1989, pp 164–8.

Talbot, A.N. and Richart, F.E. The strength of concrete: its relation to the cement aggregates and water. *Bulletin* No 137, Engineering Experiment Station, University of Illinois, 15 October, 1923.

Teychenne, D.C., Franklin, R.E. and Erntroy, H.C. Design of normal concrete mixes. Department of the Environment, 1988.

Toufar, W., Born, M. and Klose, E. Freiberger Forschungsheft A559, VEB Deutscher Verlag Fuer Grundstoffindustrie, 1967.

Toufar, W., Klose, E. and Born, M. *Aufbereitungs-technik* Vol 11, 1977, pp 603–8.

Uchikawa, H. Characterisation and material design of high-strength concrete with superior workability. *Cement Technology*, The American Ceramic Soc. 1993, pp 143–86.

Ujhelyi, J. The effect of air content of concrete, *Magyar Epitoipar* (Budapest), No. 8, 1980, pp 469–81.

Ujhelyi, J.E. Water demand of concrete mixtures. *Periodica Polytechnica Ser Civil Eng*, Vol 41, No 2, 1997, pp 199–225.

Visvesvaraya, H.C. and Mullick, A.K. Quality of cements in India – results of three decadal surveys. Uniformity of cement strength. ASTM STP 961, 1987, pp 66–79.

Walker, S. and Bloem, D.L. Effects of aggregate size on properties of concrete. NRMCA/NSGA, Joint Research Laboratory Publication No 8, September 1960, pp 283–98. Discussion Supplement 1961.

Walz, K. Herstellung von Beton nach DIN 1045, Beton-Verlag GmbH, Dusseldorf, 1971.

Wang, A., Zhang, C. and Zhang, N. Study of the particle size distribution on the properties of cement. *Cement and Concrete Research*, Vol 27, No 5, 1997, pp 685–95.

Wills, M.H. How aggregate particle shape influences concrete mixing water requirement and strength. NSGA and NRMCA Joint Research Laboratory Publication No 16, December 1967.

Wimpenny, D.E. Rheology and hydration characteristics of blended cement concretes. Sir William Halcrow and Prs, *c.* 1995, pp 1–22.

Worthington, P.M. An experiment on the theory of voids of granular materials. *MCR* April 1953, pp 121–6.

Wright, P.J.F. Entrained air in concrete. *Proceedings ICE* Part 1, Vol 2, May 1953, pp 337–58 and correspondence, pp 218–33, *c.* 1954.

Yang, M. and Jennings, H.M. Influences of mixing methods on the microstructure and rheological behaviour of cement paste. *Advanced Cement Based Materials*, 1995, 2, pp 70–8.

Further reading

Abbasi, A.F., Ahmad, M. and Wasim, M. Optimization of concrete mix proportioning using reduced factorial experimental technique. *ACI Materials Journal*, January-February 1987, pp 55–63.

ACI Guide for selecting proportions for high strength concrete with Portland cement and fly ash. Report of ACI committee 211, ACI 211.4R-93, 13 pp.

ACI. Controlled low strength materials (CLSM). ACI 229R-94 Report. *Concrete International*, July 1994, pp 55–64.

Adwan, O. Sha'at, A. and Munday, J. Mix optimization and engineering properties of high strength concrete. Dundee Institute of Technology, *c.* 1993, pp 1–15.

Ahmed, A.E. and El-Kourd, A.A. Properties of concrete incorporating natural and crushed stone very fine sand. *ACI Materials Journal*, July–August 1989, pp 417–24.

Aims, R.B. and Le Goff, P. *Powder Technology*, Vol 1, pp 281–90, 1967/8.

Allen, J.R.L. Note towards a theory of concentration of solids in natural sands. *Geological Magazine*, Vol 106, No 4, 27 August 1969, pp 309–21.

Anon. Blue Circle launches new microfine cements. *Construction Repair*, Sept/Oct 1994, p 13.

Arioglu, E. and Odbay, Discussion of Monteiro (1993). *Materials and Structures*, October 1994, Vol 27, No 172, pp 494–7.

Ayers, M.E., Wong, S.Z. and Zaman, W. Optimization of flowable fill mix proportions. ACI Special Publication SP-150, 1994, pp 16–37.

Bensted, J. An investigation of the setting of Portland cement. *Silicates Industriels*, No 6, 1980, pp 115–20.

Bloem, D.L. The problem of concrete strength relationship to maximum size of aggregate. NRMCA, Joint Research publication No 9, March 1961, pp 1–9.

Bonavetti, V.L. and Irassar, E.F. The effect of stone dust content in sand. *Cement and Concrete Research*, Vol 24, No 3, 1994, pp 580–90.

Breugel, K. van. Numerical simulation of hydration and structural development of cement-based materials. *9th Int. Conf. Chem. Cements*, New Delhi, 1993, pp 22–7.

Bruegel, K. van. Simulation model for development of properties of early age concrete. *Quality Control of Concrete Structures*. June 1991, Brussels, pp 137–52.

Clem, D.A., Hansen, K.D. and Kowalsky, J.B. Flowable backfill for pipeline bedding. ACI Special Publication SP-150, 1994, pp 87–96.

Concrete Society. Changes in the properties of ordinary Portland cement and their effects on concrete. Technical Report No 29, October 1987, 30 pp.

Concrete Society. The use of ggbs and pfa in concrete. Technical Report No 40, 1991, 142 pp.

Corish, A.T. Data on cement and concrete properties. Private correspondence with BCI plc dated March and May 1987.

Corish, A.T. Portland cement properties updated. *Concrete*, Jan/Feb 1994, pp 25–8.

Day, K.W. Equivalent slump. Production methods and workability of concrete. Paisley, 1996, pp 357–64.

de Larrard, F. and Sedran, T. Innovative approach in mixture-proportioning thanks to mathematical mixture modelling and computer technology. *ACI Fall Convention*, Montreal, Nov 1995, pp 1–6.

de Larrard, F. Batching concrete on your desk. *Concrete*, Sept 1997, pp 35–8.

Domone, P.L.J. and Soutsos, M.N. An approach to the proportioning of high-strength concrete mixes. *Concrete International*, October 1994, pp 26–31.

Drinkgern, G. Wasseranspruch sandreicher Betone. Einfluss der Sieblinie des Sandes auf die Konsistenz. Beton 9/89, pp 381–83.

Ehrenburg, D.O. Ideal and quasi-ideal grading of coarse and fine aggregates for mass concrete. *Cement, Concrete and Aggregates*, ASTM, Vol 4, No 2, Winter, 1982, pp 103–5.

Gaynor, R.D. Meininger, R.C. and Khan, T.S. Effect of temperature and delivery time on concrete proportions. NRMCA pub, No 171, June 1985, pp. 68–87.

Gaynor, R.D. Cement strength and concrete strength – An apparition or a dichotomy. *Cement, Concrete and Aggregates*, ASTM, 1993, pp 135–44.

Geymayer, H., Tritthart, J., Guo, W. and Reimann, C. Investigations on cements with different levels of fineness. *ZKG International*, 4/1995 (English), 2/1995 (German).

Hanke, V. and Krell, J. Computer aided mix design of concrete. RILEM, 1990, pp 271–6.

Hanker, V. and Siebel, E. Extended basis for concrete composition. *Beton*, 6/95 pp 412–18 and 456.

Herdan, G. *Small Particle Statistics*. Elsevier, 1953, 520 pp.

Higginson, E.C. The effect of cement fineness on concrete. Fineness of cement, ASTM Special Publication, STP 473, 1970, pp 71–81.

Hime, W.G. ASTM methods for the surface area analysis of Portland cement. Fineness of cement, ASTM Special Publication, STP 473, 1970, pp 3–19.

Hobbs, D.W. Workability and water demand. Special concretes – workability and mixing. *Proc. Int. RILEM Workshop*, March 1993.

Hughes, B.P. Some factors affecting the compressive strength of concrete. *Mag. Concr. Res.* Vol 19, No 60, Sept 1967, pp 165–72 and discussion *Mag. Concr. Res.*, Vol 20, No 63, June 1968, pp 121–2.

Hughes, B.P. and Al-Ani, M.N.A. Mix design for pfa concrete, *Proc. Instn Civ. Engrs*, Part 1, August 1990, Vol 88, pp 639–68 and June 1991, Vol 90, pp 647–8.

Hughes, B.P. and Al-Ani, M.N.A. PFA concrete mix design for the 1990s. *Cement and Concrete Composites* No 13, 1991, pp 187–95.

Johansson, A., Tuutti, K. and Petersons, N. Pumpable concrete and concrete pumping. *Advances in Ready Mixed Concrete Technology*, Pergamon Press, 1976, pp 391–404.

Kaplan, M.F. The relation between ultrasonic pulse velocity and the compressive strength of concretes having the same workability but different proportions. *Mag. Conct. Res.*, Vol 12, No 34, March 1960, pp 3–8.

Kelham, S. Active cement clinker. Part 1. Background, production and properties. 23rd annual convention ICT. 10–12 April 1995, pp 1–6.

Kilian, G. Einflussnahme uber den Mehlkorngehalt im Zuschlag auf die Betongute. *Betonwerk und Fertigteil Technik*, 11/1988, pp 75–7.

Krell, J. Influss der Feinstoffe im Beton auf die Frischbetonkonsistenz. *Neat Cement, Fresh Mortar, Fresh Concrete, Colloquium*, IBM, Hanover, 1987, pp 160–76.

Krell, J. and Wischers, G. Influss der Feinstoffe im Beton auf Konsistenz, Festigkeit und Dauerhaftigkeit. *Beton*, 9/88, pp 356–9 and 10/88, pp 401–4.

Krishnamoorthy, S. and Kantha Rao, V.V.L. Minimisation of voids content and optimization of micro filler addition in a polymer concrete mix design. *International Symposium on Innovative World of Concrete*, August/September 1993, Bangalore, India, Proceedings Vol 2, pp 253–62.

Krstulovic, P., Kamenic, N. and Popovic, K. A new approach in evaluation of filler effect in cement. II The effect of filler fineness and blending procedure. *Cement and Concrete Research*, Vol 24, No 5, 1994, pp 931–36.

Lecomte, A. and Thomas, A. Caractere fractal des melanges granulaires pour betons de haute compacite. *Materials and Structures*, No 25, 1992, pp 255–64.

Lees, G. The measurement of particle shape and its influence in engineering materials, *Jl Brit Granite and Whinstone Federation*, Vol 4, Part 2, 1964, pp 17–38.

Lees, G. Philosophy of the design of bituminous mixes for optimum structural and tyre purposes. *Journal of the Institute of Asphalt Technology*, 1985, pp 23–32.

Legrand, C. and Wirquin, E. Study of the strength of very young concrete as a function of the amount of hydrates formed-influence of super plasticizer. *Materials and Structures*, March 1994, Vol 27, No 166, pp 106–9.

Lessard, M., Gendreau, M., Baalbaki, M., Pigeon, M. and Aitcin, P-C. Formulation d'un beton a hautes performance a air entraine. *Bull. liaison Labo. P. et Ch.* 188 Nov–Dec 1993, ref 3782, pp 41–51 (In French), pp. 94–95 (Summary in English).

Loedolff, G.F. Partial replacement of cement with fly ash – optimum blends. Pretoria conference on fly ash, February 1987, pp 1–10.

Loedolff, G.F. Low porosity concrete. *RILEM Conference*, Paris, September 1987, pp 1–9.

Male, P.L. Workability and mixing of high performance microsilica concrete. Special concretes – workability and mixing. *Proc. Int. RILEM Workshop*, March 1993.

Malek, R.I.A. Effect of particle packing and mix design on chloride permeability of concrete. *9th International Congress on the Chemistry of Cement*, New Delhi, 1992, pp 143–9.

Malghan, S.G. and Lum, L-S.H. An analysis of factors affecting Particle-size Distribution of hydraulic cements. *Cement, Concrete and Aggregates*, Vol 13, No 2, Winter 1991, pp 115–20.

Monteiro, P.J.M., Helene, P.R.L. and Kang, S.H. Designing concrete mixtures for strength, elastic modulus and fracture energy. *Materials and Structures*, October 1993, Vol 26, No 162, pp. 443–52.

Murdock, L.J., Brook, K.M. and Dewar, J.D. *Concrete Materials and Practice*. Edward Arnold, 6th edition, 1991.

Newman, K. Properties of concrete. *Structural Concrete*, Vol 2, No 11, 1965, pp 451–82.

Nixon, P.J. and Spooner, D.C. Concrete proof for British cement. *Concrete*, Vol 27, No 5, Sept/Oct 1993, pp 41–4.

Numata, S. Mixture proportioning of concrete with the renewed relationship of particle interference between fine and coarse aggregates in consideration of paste stuck to their surface. *Memoirs of the Nishinippon Institute of Technology*, 26 July 1995, pp 19–30.

Parton, G.M., Shariatmadari, A.A. and Hansom, R.G. Efficiency in aggregate mix design: a least squares method. *Int. Jl of Cement Composites and Lightweight Concrete*. Vol 11, No 3, August 1969, pp 167–74.

Peris Mora, E., Paya, J. and Monzo, J. Influence of different sized fractions of a fly ash on workability of mortars. *Cem. and Conc. Res.* Vol 23, 1993, pp 917–24.

Plum, N.M. Quality control of concrete – its rational basis and economic aspects. ICE paper No 5879, pp 311–36, *c.* 1953.

Popovics, S. General relation between mix proportion and cement content of concrete. *Mag. Concr. Res.*, Vol 14, No 42, Nov 1962, pp 131–6.

Popovics, S. Contribution to the prediction of consistency. *RILEM Bulletin* No 20, September 1963, pp 91–5.

Popovics, S. An investigation of unit weight of concrete. *Mag. Concr. Res.*, Vol 16, No 49, Dec 1964, pp 211–20. Discussion MCR, Vol 17, No 52, Sept 1965.

Popovics, S. Extended model for estimating the strength-developing capacity of Portland cement. *MCR*, Vol 33, No 116, September 1981, pp 147–53.

Popovics, S. and Popovics, J.S. Computerisation of the strength versus w/c relationship. *Concrete International*, April 1995, pp 37–40.

Road Research Laboratory, DSIR Design of concrete mixes. *Road Note* No 4, 2nd edition, HMSO, 1950, pp 1–16.

Sear, L.K.A. Computer methods – Control systems. Inst. of Concrete Technology, 26th Annual Convention, Market Bosworth, UK, April 1998, pp 1–11.

Shirayama, K. The estimation of the strength of concrete made with lightweight aggregate. *Mag. Concr. Res.*, Vol 13, No 38, July 1961, pp 61–70.

Stewart, I. How to succeed in stacking. *New Scientist*, 13 July 1991, pp 29–32.

Swamy, R.N. and Lambert, G.H. Mix design and properties of concrete made from pfa coarse aggregates and sand. *Int. Journal of Cem. Composites and Lightweight Concrete*, Vol. 5, No 4, November 1983, pp 263–75.

Teychenne, D.C. The use of crushed rock aggregates in concrete. DOE/BRE, 1978, pp 74.

Tsvilis, S. and Tsimas, S. Estimation of the specific surface of the cement industry materials according to their particle size distribution. *Zement-Kalk-Gips*, No 3/1992, pp 131–4.

Tsivilis, S. and Parissakis, G. A mathematical model for the prediction of cement strength. *Cem. and Conc. Res.* Vol 25, No 1, 1995, pp 9–14.

Uchikawa, H., Hanehara, S. and Hirao, H. Cement for high-strength concrete with superior workability prepared by adjusting the composition and particle size distribution. 1995, pp 48–55.

Walter, H.-J. Mathematical model for the calculation of the optimum grain-size distribution curve with given minimum restrictions. *Betonwerk und Fertigteil Technik*, 11/1991, pp 41–5.

Wischers, G. Optimization of cements with mineral additions. *VDZ*, Dusseldorf, undated, 7 pp.

Wood, J. Experience with a computerised approach to concrete mix design. Project report for the ACT Diploma 1990/1. Institute of Concrete Technology.

Yang, C.C. and Huang, R. A two-phase model for predicting the compressive strength of concrete. *Cement and Concrete Research*, Vol 26, No 10, 1996, pp 1567–77.

Index